THE HARVEY LECTURES

DELIVERED UNDER THE AUSPICES OF

The HARVEY SOCIETY of NEW YORK

1991–1992

———

BY

STUART A. AARONSON   STANLEY B. PRUSINER

STEPHEN J. BENKOVIC   VINCENT R. RANCANIELLO

NICHOLAS R. COZZARELLI   LUBERT STRYER

STEVEN LANIER McKNIGHT   SHIRLEY M. TILGHMAN

SERIES 87

1993

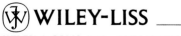 **WILEY-LISS**

A JOHN WILEY & SONS, INC. , PUBLICATION
New York • Chichester • Brisbane • Toronto • Singapore

Address All Inquiries to the Publisher
Wiley-Liss, Inc., 605 Third Avenue, New York, NY 10158-0012

Copyright © 1993 Wiley-Liss, Inc.

Printed in the United States of America

ISSN 0073-0874

ISBN 0-471-59790-2

CONTENTS

HARVEY LECTURES 1991–1992

THE HARVEY SOCIETY*

A SOCIETY FOR THE DIFFUSION OF KNOWLEDGE
OF THE MEDICAL SCIENCES

CONSTITUTION

I

This Society shall be named the Harvey Society.

II

The object of this Society shall be the diffusion of scientific knowledge in selected chapters in anatomy, physiology, pathology, bacteriology, pharmacology, and physiological and pathological chemistry, through the medium of public lectures by men and women who are workers in the subjects presented.

III

The members of the Society shall constitute two classes: Active and Honorary members. Active members shall be workers in the medical or biological sciences, residing in the metropolitan New York area, who have personally contributed to the advancement of these sciences. Active members who leave New York to reside elsewhere may retain their membership. Honorary members shall be those who have delivered lectures before the Society and who are not Active members. Honorary members shall not be eligible to office, nor shall they be entitled to a vote.

Active members shall be elected by ballot. They shall be nominated to the Executive Committee and the names of the nominees shall accompany the notice of the meeting at which the vote for their election will be taken.

IV

The management of the Society shall be vested in an Executive Committee to consist of a President, a Vice-President, a Secretary, a Treasurer, and three

*The Constitution is reprinted here for historical interest only; its essential features have been included in the Articles of Incorporation and By-Laws.

other members, these officers to be elected by ballot at each annual meeting of the Society to serve one year.

V

The Annual Meeting of the Society shall be held at a stated date in January of each year at a time and place to be determined by the Executive Committee. Special meetings may be held at such times and places as the Executive Committee may determine. At all meetings ten members shall constitute a quorum.

VI

Changes in the Constitution may be made at any meeting of the Society by a majority vote of those present after previous notification to the members in writing.

THE HARVEY SOCIETY, INC.

A SOCIETY FOR THE DIFFUSION OF KNOWLEDGE
OF THE MEDICAL SCIENCES

BY-LAWS

ARTICLE I

Name and Purposes of the Society

SECTION 1. The name of the Society as recorded in the Constitution at the time of its founding in 1905 was the Harvey Society. In 1955, it was incorporated in the State of New York as The Harvey Society, Inc.

SECTION 2. The purposes for which this Society is formed are those set forth in its original Constitution and modified in its Certificate of Incorporation as from time to time amended. The purposes of the Society shall be to foster the diffusion of scientific knowledge in selected chapters of the biological sciences and related areas of knowledge through the medium of public delivery and printed publication of lectures by men and women who are workers in the subjects presented, and to promote the develeopment of these sciences.

It is not organized for pecuniary profit, and no part of the net earnings, contributions, or other corporate funds of the Society shall inure to the benefit of any private member or individual, and no substantial part of its activities shall be carrying on propaganda, or otherwise attempting, to influence legislation.

ARTICLE II

Offices of the Society

SECTION 1. The main office and place of business of the Society shall be in the City and County of New York. The Board of Directors may designate additional offices.

ARTICLE III

Members

SECTION 1. The members of the Society shall consist of the incorporators, members of the hitherto unincorporated Harvey Society, and persons elected from time to time. The members of the Society shall constitute two classes: Active and Honorary Members. Active members shall be individuals with either the Ph.D. or the M.D. degree or its equivalent, residing or carrying on a major part of their work in the New York metropolitan area at the time of their election, who are personally making original contributions to the literature of the medical or biological sciences. Honorary members shall be those who have delivered a lecture before the Society and who are not Active members. Honorary members shall be exempted from the payment of dues. Active members who have remained in good standing for 35 years or who have reached the age of 65 and have remained in good standing for 25 years shall be designated Life members. They shall retain all the privileges of their class of membership without further payment of dues. Honorary members shall not be eligible to office, nor shall they be entitled to participate by voting in the affairs of the society. Volumes of The Harvey Lectures will be circulated only to Active members. Life members will be offered the opportunity to purchase The Harvey Lectures at the cost of the volume. Honorary members will receive only the volume containing their lecture. New Active members shall be nominated in writing to the Board of Directors by an Active member and seconded by another Active member. They shall be elected at the Annual Meeting of the Society by a vote of the majority of the Active members present at the meeting. Members who leave New York to reside elsewhere may retain their membership. Active members who have given a Harvey Lecture and who have moved out of the New York metropolitan area may, if they wish, become Honorary members. Membership in the Society shall terminate on the death, resignation, or removal of the member.

SECTION 2. Members may be suspended or expelled from the Society by the vote of a majority of the members present at any meeting of members at which a quorum is present, for refusing or failing to comply with the By-Laws, or for other good and sufficient cause.

SECTION 3. Members may resign from the Society by written declaration, which shall take effect upon the filing thereof with the Secretary.

ARTICLE IV

Meetings of the Members of the Society

SECTION 1. The Society shall hold its annual meeting of Active members for the election of officers and directors, and for the transaction of such other business as may come before the meeting in the month of January or

February in each year, at a place within the City of New York, and on a date and at an hour to be specified in the notice of such meeting.

SECTION 2. Special meetings of members shall be called by the Secretary upon the request of the President or Vice-President or of the Board of Directors, or on written request of twenty-five of the Active members.

SECTION 3. Notice of all meetings of Active members shall be mailed or delivered personally to each member not less than ten nor more than sixty days before the meeting. Like notice shall be given with respect to lectures.

SECTION 4. At all meetings of Active members of the Society ten Active members, present in person, shall constitute a quorum, but less than a quorum shall have power to adjourn from time to time until a quorum be present.

ARTICLE V

Board of Directors

SECTION 1. The number of directors constituting The Board of Directors shall be seven: the President, the Vice-President, the Secretary, and the Treasurer of the Society, and the four members of the Council. The number of directors may be increased or reduced by amendments of the By-Laws as hereinafter provided, within the maximum and minimum numbers fixed in the Certificate of Incorporation or any amendment thereto.

SECTION 2. The Board of Directors shall hold an annual meeting shortly before the annual meeting of the Society.

Special meetings of the Board of Directors shall be called at any time by the Secretary upon the request of the President or Vice-President or of one-fourth of the directors then in office.

SECTION 3. Notice of all regular annual meetings of the Board shall be given to each director at least seven days before the meeting and notice of special meetings, at least one day before. Meetings may be held at any place within the City of New York designated in the notice of the meeting.

SECTION 4. The Board of Directors shall have the immediate charge, management, and control of the activities and affairs of the Society, and it shall have full power, in the intervals between the annual meetings of the Active members, to do any and all things in relation to the affairs of the Society.

SECTION 5. Council members shall be elected by the members of the Society at the Annual Meeting. One Council member is elected each year to serve for three years, there being three Council members at all times. Vacancies occurring on the Council for any cause may be filled for the unexpired term by the majority vote of the directors present at any meeting at which a quorum is present. Only Active members of the Society shall be eligible for membership on the Council.

SECTION 6. A majority of the Board as from time to time constituted shall be necessary to constitute a quorum, but less than a quorum shall have power to adjourn from time to time until a quorum be present.

SECTION 7. The Board shall have power to appoint individual or corporate trustees and their successors of any or all of the property of the Society, and to confer upon them such of the powers, duties, or obligations of the directors in relation to the care, custody, or management of such property as may be deemed advisable.

SECTION 8. The directors shall present at the Annual Meeting a report, verified by the President and Treasurer, or by a majority of the directors, showing the whole amount of real and personal property owned by the Society, where located, and where and how invested, the amount and nature of the property acquired during the year immediately preceding the date of the report and the manner of the acquisition; the amount applied, appropriated, or expended during the year immediately preceding such date, and the purposes, objects, or persons to or for which such applications, appropriations, or expenditures have been made; and the names of the persons who have been admitted to membership in the Society during such year, which report shall be filed with the records of the Society and an abstract thereof entered in the minutes of the proceedings of the Annual Meeting.

ARTICLE VI

Committees

SECTION 1. The Board of Directors may appoint from time to time such committees as it deems advisable, and each such committee shall exercise such powers and perform such duties as may be conferred upon it by the Board of Directors subject to its continuing direction and control.

ARTICLE VII

Officers

SECTION 1. The officers of the Society shall consist of a President, a Vice-President, a Secretary, and a Treasurer, and such other officers as the Board of Directors may from time to time determine. All of the officers of the Society shall be members of the Board of Directors.

SECTION 2. The President shall be the chief executive officer of the Society and shall be in charge of the direction of its affairs, acting with the advice of the Board of Directors. The other officers of the Society shall have the powers and perform the duties that usually pertain to their respective offices, or as may from time to time be prescribed by the Board of Directors.

SECTION 3. The officers and the directors shall not receive, directly or indirectly, any salary or other compensation from the Society, unless authorized by the concurring vote of two-thirds of all the directors.

SECTION 4. The officers shall be elected at the Annual Meeting of the Active members. All officers shall hold office until the next Annual Meeting and until their successors are elected or until removed by vote of a majority of the directors. Vacancies occurring among the officers for any cause may be filled for the unexpired term by the majority vote of the directors present at any meeting at which a quorum is present. Officers must be Active members of the Society.

ARTICLE VIII

Fiscal Year—Seal

SECTION 1. The fiscal year of the Society shall be the calendar year.

SECTION 2. The seal of the Society shall be circular in form and shall bear the words "The Harvey Society, Inc., New York, New York, Corporate Seal."

ARTICLE IX

Amendments

SECTION 1. These By-Laws may be added to, amended, or repealed, in whole or in part, by the Active members or by the Board of Directors, in each case by a majority vote at any meeting at which a quorum is present, provided that notice of the proposed addition, amendment, or repeal has been given to each member or director, as the case may be, in the notice of such meeting.

OFFICERS OF THE HARVEY SOCIETY

1991–1992

KENNETH I. BERNS, *President*
JERARD HURWITZ, *Vice President*
JOSEPH R. BERTINO, *Treasurer*
PETER PALESE, *Secretary*

COUNCIL

1991–1992

FRIENDS OF THE HARVEY SOCIETY

1991–1992

The Harvey Society gratefully acknowledges the support of the following corporate "Friends" whose generosity has contributed to the success of this Lecture Series.

The Dupont Merck Pharmaceutical Company

CONTRIBUTORS TO THE HARVEY SOCIETY

1991–1992

Amgen, Inc.
Hoffmann-LaRoche, Inc.
Miles, Inc./Bayer AG

VIRUS–RECEPTOR INTERACTION IN POLIOVIRUS ENTRY AND PATHOGENESIS

VINCENT R. RACANIELLO

Department of Microbiology, Columbia University College of Physicians and Surgeons, New York, New York

I. Introduction

THE initial interactions between a virus and a host cell play a crucial role in viral infections. Upon encountering a target cell, the capsid binds to a cell surface receptor, initiating events that lead to release of the viral nucleic acid into the cell. The same capsid also protects the viral genome as it travels between cell targets. The viral capsid must, therefore, be stable to the extracellular environment, yet flexible enough to discharge the nucleic acid into the cell. Knowledge of how the viral capsid and the cell receptor mediate these steps is clearly required to fully understand viral replication and pathogenesis. Poliovirus is a particularly good model for studying early events in virus infection. The three-dimensional structure of poliovirus has been determined by X-ray crystallography (1), the cell receptor for poliovirus has been identified (2), and genetic manipulation of the virus is possible with infectious cDNA copies of the viral genome (3).

Poliovirus is a member of the Picornaviridae, a virus family that includes the enteroviruses (e.g., polioviruses, coxsackieviruses, echoviruses), hepatovirus (hepatis A virus), rhinovirus, aphthovirus (foot and mouth disease virus) and cardiovirus (e.g., encephalomyocarditis virus and mengovirus). These viruses are responsible for a variety of human and animal diseases. The capsids of picornaviruses are composed of 60 copies each of proteins VP1, VP2, VP3, and VP4, arranged in icosahedral symmetry around a single-stranded message-sense RNA genome (Fig. 3). Polioviruses are grouped into three serotypes based on antigenicity of the capsid.

1

The Harvey Lectures, Series 87, pages 1–16

II. Early Events In Poliovirus Infection

Poliovirus replication begins when the virus binds to a cell surface receptor, which has been identified as a new member of the immunoglobulin superfamily (2). The poliovirus receptor (PVR) consists of three extracellular immunoglobulin (Ig-like) domains, a transmembrane domain, and a cytoplasmic tail (Fig. 1). Alternative splicing results in two mRNAs that encode polypeptides of 43 and 45 kd with different cytoplasmic domains. Secreted PVR isoforms lacking the transmembrane domain are also produced by alternative mRNA splicing (4). PVR is the receptor for all three serotypes of poliovirus, and is distinct from receptors used by other picornaviruses (5).

Mouse cell cultures are not susceptible to poliovirus infection because these cells do not express PVR (6). However, expression of PVR in mouse cells from cloned cDNA results in susceptibility to poliovirus infection (2). Similarly, mice are not susceptible to most strains of poliovirus, and when PVR

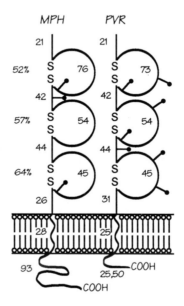

Fig. 1.   Predicted structures of MPH and PVR proteins. The Ig domain is schematically represented as a loop joined by a disulfide bond. Lengths in amino acids, and percent amino acid identity for domains 1, 2, and 3 are shown. Ball and stick show predicted N-linked glycosylation sites.

is expressed in the germline of mice, the resulting transgenic animals become susceptible to poliovirus infection (7). PVR is therefore the determinant of poliovirus host range in mice.

After binding to cell surface PVR, poliovirus must release its RNA genome into the cell. The mechanism of this uncoating step remains an enigma. If poliovirus is bound to cells at temperatures below 33°C, the attached viruses can be released in an infectious form by exposure to 6M LiCl, 8M urea, low pH, or detergents (8). When poliovirus is bound to cells at 37°C, a substantial fraction of the bound viruses release their copies of VP4 and are converted into a conformationally altered form known as the "A" particle (9). "A" particles sediment more slowly than native viruses, are noninfectious (although they contain infectious RNA), and are sensitive to detergent and proteinases. The N-terminus of VP1, which is on the interior of the native particle, is on the surface of "A" particles (10).

Experimental evidence suggests that "A" particles release their RNA into the cytoplasm, although this uncoating process is not well understood. "A" particles can be found within cells early after infection (11). They are more hydrophobic than native virions as a result of exposure of the amino terminus of VP1, the first 20 residues of which resemble an amphipathic helix (10). It has been suggested that five copies of the VP1 N-terminus might create a pore in the cell membrane that could lead to passage of the viral genome into the cytoplasm (10). Consistent with this hypothesis is the observation that a mutant virus lacking VP1 amino acids 8 and 9 is defective in uncoating (12). Because the infectivity of typical preparations of poliovirus is often only 0.5% of the physical population, it is difficult to assess the role of intermediates such as "A" particles in infectivity. However, the finding that the antiviral WIN (Sterling-Winthrop, Inc.) compounds block the formation of "A" particles strongly suggests an important role for these particles in infection (13).

A model for the early stages of poliovirus entry is presented in Figure 2. Virus attaches to cell surface PVR and undergoes a receptor-mediated conformational transition to "A" particles. While many "A" particles elute from the cell, a small fraction remains bound and releases RNA into the cell. The trigger for this uncoating step is not known; it might result from multivalent binding to PVR, or it might require the low pH environment that occurs during endocytosis. Poliovirus particles can be seen in coated pits and endosomes shortly after adsorption, suggesting that entry occurs by receptor- mediated endocytosis (14,15). Whether or not acidification of endosomes is required for uncoating is a matter of controversy. It has been reported that

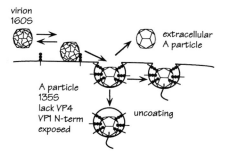

Fig. 2. Model for poliovirus entry. Binding is reversible when carried out at temperatures below 33°C; at higher temperatures, bound viruses are converted to "A" particles. "A" particles are found both outside and inside the cell, and may be entry intermediates. Uncoating of the "A" particle may occur in "pits," or from within endosomes.

raising the pH after adsorption with monensin or with weak bases inhibits uncoating (15,16). More recently it has been suggested that elevation of endosomal pH does not affect poliovirus uncoating (17).

Analogies with the conformational transitions that are believed to occur during cell entry of plant viruses have lead to the suggestion that the interfaces between protomers are important in controlling structural transitions that occur during uncoating (18). The protomer is a subunit of the capsid, which in poliovirus consists of a single copy of each of the four virion polypeptides (Fig. 3). Upon exposure to chelators of divalent cations at alkaline pH, many icosahedral plant viruses undergo an expansion in radius (19). This expansion is regulated by an interface between capsid proteins that contains a pair of divalent cation binding sites. When divalent cations are removed, this protomer interface is destabilized. The resulting expansion of the particle also causes externalization of the amino terminus of the capsid proteins. The atomic structure of the expanded form of tomato bushy stunt virus reveals that the amino termini are extruded through a gap produced by disruption of the protomer interface. In poliovirus, which is structurally similar to these plant viruses, the N-terminus of VP1 is located directly below the analogous interface. The receptor-mediated conformational transition of poliovirus may therefore involve disruption of this interface to permit extrusion of the VP1 N-terminus.

Another site that is believed to modulate receptor-mediated conformational transitions in poliovirus is a hydrocarbon-binding pocket in VP1, which is

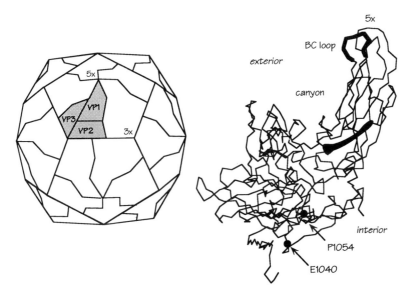

Fig. 3.   Locations of the VP1 B–C loop and N-terminal suppressors that determine host range. The virion is shown at left, with one protomer highlighted. At right is an α-carbon trace of VP1 and VP2 only. The B–C loop is highlighted at the fivefold axis, and the sites of the N-terminal suppressors are shown. Sphingosine in the hydrocarbon-binding pocket is shown in gray.

normally occupied by sphingosine. Antiviral drugs of the WIN class bind to the analogous pocket in human rhinovirus 14 (HRV14) and prevent attachment to cells. These drugs do not prevent attachment of poliovirus to cells, but block the formation of ''A'' particles (13). The presence of drug in the pocket might make the capsid more rigid, preventing the receptor-mediated conformational rearrangements that occur during cell entry. Precisely how the drugs effect this change is not known. One possibility is that the antiviral drugs displace amino acid residues that control conformational transitions. These drugs have been shown to displace HRV14 residues in the canyon floor and at the protomer interface, which lies just above the hydrocarbon-binding pocket (20). WIN compounds also stabilize the virion against thermal inactivation (21), further demonstrating the role of the hydrocarbon-binding pocket in controlling structural transitions.

III. Capsid Sequences That Control Interaction Of Poliovirus With PVR

Resolution of the crystal structures of poliovirus type 1 (1) and HRV14 (22) has provided insight into how these viruses might recognize their cell receptors. It has been proposed that a deep depression on the surface of these viruses, called the canyon, is the receptor binding site (23). The canyon encircles each fivefold axis of symmetry on the virion particle. Several lines of evidence support the hypothesis that the canyon is the receptor binding site. Site-directed mutagenesis of amino acid residues on the walls and floor of the HRV14 canyon produced viruses with altered binding affinities to cells (24). The antiviral WIN compounds bind in the hydrophobic pocket beneath the canyon floor, leading to conformational changes in the floor that are believed to interfere with the binding of HRV14 to its cell receptor (20). Low-resolution images of HRV1G bound to its soluble cell receptor, ICAM-1, show ICAM-1 inserted into the canyon (M. Rossmann, personal communication).

Because of the structural similarity between poliovirus and rhinovirus, the canyon of poliovirus might be the binding site for PVR. However, no experimental information is available concerning how poliovirus attaches to PVR. Our approach to this problem has been to study poliovirus mutants selected for their resistance to neutralization with soluble PVR (25). Incubation of poliovirus at 37°C with detergent extracts of insect cells expressing PVR (called S100PVR) results in neutralization of viral infectivity. Studies on the mechanism of neutralization revealed that poliovirus is converted to 135S "A" particles after incubation with S100PVR (26). Mutants of poliovirus type 1 Mahoney could be selected for resistance to soluble PVR, with a frequency of 10^{-4}–10^{-5}, comparable to that reported for isolation of mutants resistant to monoclonal antibodies (mAbs), and the antiviral WIN compounds (27,28). These soluble receptor resistant (*srr*) mutants use PVR to enter and replicate in HeLa cells.

The mutation responsible for the *srr* phenotype has been identified in 20 independent isolates of the P1/Mahoney strain (unpublished results). Fourteen different amino acid changes were observed at 11 capsid positions, 9 in VP1 and 2 in VP3. To obtain clues about how these mutations confer the *srr* phenotype, their locations in the three-dimensional structure of the capsid were determined. Eight sites of mutation are located on the virion surface and 3 are on the interior of the virion. All the surface mutations are located in the canyon, near the interface between protomers. The mutations on the interior of the capsid are located near the hydrocarbon-binding pocket in VP1.

The results of competition binding assays indicate that some of the *srr* mutants have reduced binding affinity for HeLa cells. The locations of the responsible mutations in the poliovirus capsid provide clues about how they lead to the binding defect. The surface mutations might occur at sites that contact PVR. Mutations at these positions could influence receptor contact and reduce binding affinity.

Another explanation for how mutations in *srr* variants affect binding is suggested by the observation that all the surface mutations are located at the interface between fivefold related protomers. As discussed previously, residues that surround this interface are believed to regulate receptor-mediated structural transitions in poliovirus (18). Because of the large area of potential contact between virus and receptor, it may be advantageous to have flexibility in the shape of the virus, to allow it to maximize contact with PVR. Because the protomer interface controls conformational transitions in the particle, it may also regulate transitions that affect binding affinity. The requirement for conformational changes to produce high-affinity binding is also believed to occur in antigen-antibody reactions (29).

A similar explanation could account for the effect of the changes at internal positions on binding affinity. These residues surround the hydrocarbon-binding pocket in VP1. The pocket residues are also believed to regulate conformational transitions of the capsid. As a result of their vicinity to the pocket, these *srr* mutations might influence structural transitions, thereby affecting binding affinity.

IV. PVR SEQUENCES THAT CONTROL POLIOVIRUS INFECTION

Another interesting aspect of the virus–receptor interaction is the identification of structural features of the receptor required for infection of cells. Several approaches have been used to identify viral binding sites for human immunodeficiency virus-1 (HIV-1) and the major group human rhinoviruses (HRV) on CD4 and ICAM-1, respectively, including deletion mutagenesis and construction of chimeras between the human receptors and murine homologs that do not bind virus. The first Ig-like domain of CD4 contains a high-affinity binding site for the HIV-1 glycoprotein gp120 (30,31), while the first of the five Ig-like domains of ICAM-1 contains the binding site for IRV (32–34).

The first Ig-like domain of PVR (a V-type domain) contains the binding site for poliovirus. Because the atomic structure of PVR has not been determined, the Ig-like domains (domain 1, amino acids (aa) 35–142; domain 2, aa 153–236; domain 3, aa 250–330) can be modeled after known immunoglobulin structures (35–37). Cells expressing PVR lacking aa 33–137 or 40–136 cannot bind poliovirus or support virus infection (38–40). Expression of PVR aa 1–164 linked to the transmembrane domain is sufficient to confer susceptibility to poliovirus infection; however, this polypeptide encodes 28 aa beyond the predicted end of domain 1 (38,39). This additional sequence may be required for poliovirus infection, because deletion of aa 137–256 destroys receptor activity (40). Thus, domain 1 and an additional 28 aa are the minimum structure needed to confer sensitivity to poliovirus infection.

The construction and analysis of chimeric proteins between human receptors and their murine homologues have provided much information about the binding sites for HIV-1 and HRV on CD4 and ICAM-1. A *m*urine *P*VR *h*omolog, MPH, has also been identified (41). Amino acid sequence alignment of MPH and PVR indicates conservation of the three extracellular Ig-like domains across species (Fig. 1). The three domains (defined by intron-exon boundaries) share 52%, 57%, and 64% identity and 64%, 68% and 80% similarity at the amino acid level. While MPH does not bind poliovirus, when aa sequence 1–140 (including the first Ig-like domain) of MPH is substituted with the corresponding sequence from PVR, the resulting chimeric protein can support poliovirus infection. Therefore the amino acid differences in domain 1 between MPH and PVR prevent MPH from functioning as a poliovirus receptor. Because PVR and MPH share extensive homology in domain 1 (52% identity, 64% similarity), it should be possible, by substitution of murine MPH for PVR residues in the individual loops and β-strands of the first immunoglobulin-like domain, to identify PVR residues critical for poliovirus binding and infection.

V. PVR AND POLIOVIRUS HOST RANGE

The natural hosts for polioviruses are humans, but chimpanzees and certain species of monkeys can be experimentally infected, and strains of polioviruses have been identified that cause paralysis in mice. Intracerebral or intraspinal inoculation of mice with these strains leads to a disease that resembles human poliomyelitis in histopathology and in age-dependence of susceptibility (42,43). The virulence of the P2/Lansing strain was acquired after a process of adaptation involving serial passage of viruses in mice (44). A number of viral strains, such as the P1/Mahoney strain, are host restricted and cause paralysis in primates but not in mice. When the gene encoding the human poliovirus receptor is introduced into the germline of mice, the resulting transgenic animals can be infected by P1/Mahoney (7). Therefore, the primary block to infection of normal mice by these strains is at the level of cell entry. Thus, molecular analysis of poliovirus host range provides a way of investigating virus–receptor interactions and may shed light on how poliovirus enters cells.

Construction of viral recombinants between the P2/Lansing strain and the host-restricted P1/Mahoney, which causes paralysis in primates but not in mice, revealed that the capsid coding region of P2/Lansing is sufficient to confer the mouse virulent phenotype on the P1/Mahoney strain (45). Monoclonal antibody (mAb) resistant variants of P2/Lansing, containing mutations in neutralization antigenic site 1 (N-Ag1) have reduced mouse neurovirulence (46). N-Ag1 is one of three sites on the capsid surface that together define the three antigenic types of poliovirus. Amino acids 95–104 of capsid protein VP1, which contributes substantially to N-Ag1, comprise a loop connecting β-strands B and C (the B–C loop) of VP1. This loop is highly exposed on the virion surface near the fivefold axis of symmetry (Fig. 3). When the VP1 B–C loop of P1/Mahoney is substituted with the sequence from P2/Lansing, the resulting chimeric virus is virulent in mice (47,48). The reciprocal recombinant is attenuated in normal mice and virulent in PVR transgenic mice (49). Therefore the B–C loop of VP1 is the primary determinant of the host restriction of P1/Mahoney and the mouse virulence of P2/Lansing.

Two amino acid changes that suppress the requirement for the P2/Lansing-specific B–C loop sequence for infection of mice have been identified (49). These suppressor mutations are located in the N-terminal extension of VP1, on the interior surface of the virus particle. Like the VP1 B–C loop, each of the two mutations can extend the host range of poliovirus to include mice. The B–C loop and the internal suppressors are therefore host range determinants.

Host restriction of poliovirus P1/Mahoney in mice can therefore be overcome in three ways: by substituting the P1/Mahoney-specific VP1 B–C loop with that of P2/Lansing, by introducing specific intragenic suppressors into the N-terminus of VP1, or by expressing PVR in transgenic mice. Because the human receptor can overcome host restriction, it seems likely that polioviruses which cannot infect mice are blocked in one of the events of the virus–receptor interaction. Our hypothesis is that P2/Lansing is able to recognize receptors in normal mice that cannot be recognized by P1/Mahoney. How then do the B–C loop and the N-terminus of VP1 control this receptor recognition? The B–C loop, located above the north wall of the canyon (Fig. 3), could be part of the receptor recognition site. Because P2/Lansing variants lacking the B–C loop can replicate in cultured human cells and in PVR transgenic mice, this site must be different from that used by the virus to recognize the human receptor. However, when attenuation suppressors in the N-terminus of VP1 are present, the B–C loop sequences required for mouse virulence are significantly different. It is not clear how internal mutations overcome host restriction if the B–C loop is a receptor recognition site. One possibility is that changes in the N-terminus of VP1 affect the conformation of the B–C loop.

A more attractive explanation for the mechanism of action of the internal host range determinants is that they affect the ability of the virion to undergo structural changes early in infection. The internal suppressors in VP1, gly-40 and ser-54 (Fig. 3), are located near sites that may regulate such transitions, including the seven-stranded β-sheet, and the interface between fivefold-related protomers (18). The B–C loop may also determine host range by affecting conformational transitions of the virion. The B–C loop might influence the structure or movement of the E–F loop, which forms part of the interface between fivefold-related protomers. The E–F loop is part of N-Ag1, which also consists of the D–E, H–I, and B–C loops, and thus clearly interacts structurally with the other loops. Structure analysis of a P1/Mahoney chimera with a P2/Lansing B–C loop demonstrates that the B–C loop structurally interacts with other loops at the fivefold axis (50). Interactions along the protomer interface are likely to play significant roles in the dynamics of the capsid associated with receptor attachment and cell entry. By influencing such interactions, the B–C loop and the N-terminus of VP1 might cooperatively regulate conformational transitions in the capsid. If the virus–receptor interaction is concurrent with early conformational transitions in the virus particle, then different receptors or capsid residues could influence the ability of the virion to undergo such transitions, thereby determining the outcome of infection. Productive interaction between the virus and the mouse receptor may have slightly different requirements than the interaction with the human receptor. As a result, P1/Mahoney might effectively use only the human receptor while P2/Lansing virus might use both receptors. Answers to these questions will require identification of the cell receptor that P2/Lansing uses to infect mice.

VI. TgPVR Mice and the Pathogenesis of Poliomyelitis

In addition to its narrow host range, poliovirus has a restricted cell and tissue tropism. Although poliovirus infection includes a viremic stage during which virus has access to all tissues, viral replication is largely confined to the oropharyngeal and intestinal mucosa, neurons within the central nervous system, and an unidentified extraneural site (reviewed in ref. 51). Furthermore, poliovirus does not replicate when inoculated directly into monkey kidney or testicular tissue (52–54). A simple explanation for this restricted tissue tropism is that it is determined by receptor distribution. Early approaches to this question involved examining the ability of poliovirus to bind homogenates of different tissues. This assay depends on the loss of infectivity associated with receptor binding at 25°C, which presumably occurs when PVR blocks attachment sites on the virion. In one study it was found that poliovirus bound to homogenates of susceptible tissues but not to those of nonsusceptible tissues (55). However, occasional low levels of virus binding to kidney, liver, and lung were also observed. In another study, significant virus binding to some monkey and human tissues, including those that do not support virus replication, was reported (56), suggesting that some nonsusceptible tissues also express receptors. In still another study, the binding of radiolabeled poliovirus to human regional CNS tissue homogenates showed that the binding activity within the CNS is much more widespread than the distribution of virus-induced pathologic lesions (57). These results illustrate the difficulty in determining PVR presence by virus binding assays with tissue homogenates. Such assays might not detect low levels of binding in certain tissues, especially if PVR is degraded by tissue proteinases.

The availability of PVR cDNAs enables further examination of the relationship between poliovirus tissue tropism and expression of cell receptors. Northern and western blot analyses indicate that PVR RNA and protein are expressed in many human tissues, including those that do not support poliovirus replication (7); this result confirms that there is no simple correlation between tropism and expression of PVR RNA and protein. There are many ways to reconcile these results with those obtained from virus binding assays conducted 40 years ago. While PVR protein is expressed in many tissues, it might not always find its way to the cell surface. Alternatively, the ability of cell surface PVR to bind poliovirus might be controlled by post-translational modifications. Because virus binding assays might not adequately detect PVR at the cell surface, it is possible that all tissues express PVR that is able to bind poliovirus, and that tropism is determined at a postbinding step. Clearly more information on the subcellular location of PVR protein in human tissues, and its ability to bind poliovirus is required to address these issues.

The availability of PVR transgenic mice susceptible to poliovirus infection has enabled a study of the determinants of poliovirus tissue tropism in an easily manipulatable animal model (7). PVR RNA is expressed in all organs of TgPVR mice, in a cell-type–specific manner that is independent of transgene insertion site or copy number (58). Most neurons in all areas of the CNS and PNS express high levels of PVR transcripts. In the kidney, high levels of PVR transcripts were detected in renal corpuscles and in some tubular epithelial cells in the medulla, but not in the renal pelvis, nor in lymph nodes, fatty tissue, or blood vessels that surround the kidney. In the lung, PVR transcripts were detected in cells, tentatively identified as macrophages, lining the alveoli. Bronchial epithelial cells expressed lower levels of PVR transcripts. PVR RNAs were also detected in T-lymphocytes in the cortex of the thymus, as well as in some cells in the medulla of the thymus, and in endocrine cells of the adrenal cortex. PVR gene expression was detected at low levels in most cells of the intestine, the spleen, and skeletal muscle.

These experiments do not indicate whether PVR protein is expressed in TgPVR cells. Homogenates of TgPVR brain, intestine, liver, lung, and kidney bound significant levels of poliovirus, and no binding activity was detected in tissue homogenates of nontransgenic mice. To determine whether PVR protein is displayed on the cell surface, TgPVR kidney was dispersed and assayed for its ability to bind poliovirus. Cultured TgPVR kidney cells bind poliovirus, indicating that there is cell surface expression in at least some TgPVR kidney cells. Whether or not other cell types express poliovirus binding sites at the surface remains to be determined.

Despite this widespread pattern of PVR RNA expression, poliovirus replication in TgPVR mice is restricted to few sites: neurons of the CNS, skeletal muscle, and brown adipose tissue. Cultured TgPVR kidney cells, which bind poliovirus, are resistant to poliovirus infection, indicating that in this organ, the block to poliovirus replication is not at the level of binding. One or more subsequent events in the viral replicative cycle, such as penetration, uncoating, or biosynthesis, could be obstacles to poliovirus replication in TgPVR kidney. The restriction of poliovirus replication in other tissues may be different.

Poliovirus entry might require factors in addition to PVR that are lacking in nonsusceptible tissues. For example, expression of human CD4 in rodent cells is not sufficient to render these cells susceptible to HIV-1 infection, due to a block at the level of entry (59). The tissue expression pattern of the

100 kd glycoprotein recognized by mAb AF3 is strikingly similar to that observed for poliovirus susceptibility (60). While it is tempting to suggest that the 100 kd protein is in some way involved in controlling poliovirus tropism, this possibility seems unlikely. Expression of PVR in mouse cells is sufficient to confer susceptibility to poliovirus infection, yet mouse cells do not express the 100 kd glycoprotein (60).

TgPVR mice have also been useful for studying the route by which poliovirus reaches the CNS (61). Two possibilities have been widely debated: Virus enters the CNS from the blood across the blood-brain barrier (BBB), or virus enters a peripheral nerve and is transmitted to the CNS (62–68). In PVR transgenic mice, the LD_{50} for intramuscular and intracerebral inoculation is similar, suggesting that poliovirus may reach the CNS directly after intramuscular inoculation. Paralysis is initially observed in the injected limb in transgenic mice, but not in nontransgenic mice inoculated intramuscularly with P2/Lansing, a mouse-adapted poliovirus strain. Following intramuscular injection, poliovirus spreads first to the inferior segment of the spinal cord, then to the superior spinal cord, and then the brain. Finally, development of CNS disease after inoculation in the hindlimb footpad is blocked by sciatic nerve transection. These results directly demonstrate that polioviruses spread from muscle to the CNS through nerve pathways. A similar mechanism may be responsible for the spread of poliovirus in humans (61).

VII. SUMMARY

Knowledge of the three-dimensional structure of poliovirus has provided insight into many aspects of poliovirus infection. However, the structure alone cannot answer questions about how the virus interacts with the receptor to lead to release of the genome into the cell. Genetic and biochemical analysis of virus mutants are required to understand these steps in infection. The availability of cDNA clones of the poliovirus genome and PVR enables the isolation of virus and PVR mutants that can be used to study early events in infection. The results of these studies, coupled with the resolution of the three-dimensional structure of PVR, and perhaps of a PVR–poliovirus complex, should provide a detailed picture of virus binding, alteration, and uncoating.

Viral receptors do not exist solely for the benefit of viruses, but serve a particular function in the cell. The study of virus receptors, therefore, bridges the disciplines of virology and cell biology. Identification of viral receptors may reveal previously unknown cell proteins whose functions can be studied, or might enhance our knowledge of a known protein. As a member of the Ig superfamily of proteins, PVR is likely to have a role in cell adhesion and/or cell–cell communication. It will, therefore, be important to identify a PVR ligand or coreceptor. The study of the expression of PVR and its ligand will provide a basis for understanding the normal function of this protein. Determination of the function of MPH will be facilitated by the analysis of mice containing a targeted disruption of the MPH gene. It is likely that knowledge of the cell function of MPH can also provide information on the interaction of poliovirus with host cells. It has been suggested that virus binding to cell receptors may lead to activation of cell events that lead to disease (69). While there is no evidence that poliovirus affects cells in this way, studying the interaction of MPH with its ligand might provide clues about cell processes that may be activated by poliovirus infection.

Acknowledgments

I am grateful to the talented graduate students, postdoctoral fellows, and technicians who have worked in my laboratory to obtain the results described in this chapter. I am also thankful for the support of our research by the National Institutes of Health, the American Cancer Society, Lederle-Praxis Laboratories, and the World Health Organization.

References

1. Hogle, J.M., Chow, M., and Filman, D.J. (1985). *Science* **229**, 1358–1365.
2. Mendelsohn, C., Wimmer, E., and Racaniello, V.R. (1989). *Cell* **56**, 855–865.
3. Racaniello, V.R., and Baltimore, D. (1981). *Science* **214**, 916–919.
4. Koike, S., Horie, H., Dise, I., Okitsu, H., Yoshida, M., Iizuka, N., Takeuthi, K., Takegami, T., and Nomoto, A. (1990). *EMBO J.* **9**, 3217–3224.
5. Colonno, R.J. (1986). *Bioessays* **5**, 270–274.
6. McLaren, L.C., Holland, J.J., and Syverton, J.T. (1959). *J. Exp. Med.* **109**, 475–485.
7. Ren, R., Costantini, F.C., Gorgacz, E.J., Lee, J.J., and Racaniello, V.R. (1990). *Cell* **63**, 353–362.
8. Lonberg-Holm, K., and Philipson, L. (1974). *Monogr. Virol.* **9**, 1–148.
9. Joklik, W.K., and Darnell, J.E. (1961). *Virology* **13**, 439–447.
10. Fricks, C.E., and Hogle, J.M. (1990). *J. Virol.* **64**, 1934–1945.
11. Everaert, L., Vrijsen, R., and Boeyé, A. (1989). *Virology* **171**, 76–82.
12. Kirkegaard, K. (1990). *J. Virol.* **64**, 195–206.
13. Fox, M.P., Otto, M.J., and McKinlay, M.A. (1986). *Antimicrob. Agents Chemother.* **30**, 110–116.
14. Willingmann, P., Barnert, H., Zeichhardt, H., and Habermehl, K.-O. (1989). *Virology* **168**, 417–420.

15. Zeichhardt, H., Wetz, K., Willingmann, P., and Habermehl, K.-O. (1985). *J. Gen. Virol.* **66,** 483–492.
16. Madshus, I.H., Olsnes, S., and Sandvig, K. (1984). *J. Cell. Biol.* **98,** 1194–1200.
17. Gromeier, M., and Wetz, K. (1990) *Virology* **64,** 3590–3597.
18. Filman, D.J., Syed, R., Chow, M., Macadam, A.J., Minor, P.D., and Hogle, J.M. (1989). *EMBO J.* **8,** 1567–1579.
19. Incardona, N.L., and Kaesberg P. (1974). *Biophys.* **4,** 11–21.
20. Badger, J., Minor, I., Kremer, M.J., Oliveira, M.A., Smith, T.J., Griffith, J.P., Guerin, D.M.A., Krishnaswamy, S., Luo, M., Rossmann, M.G., McKinlay, M.A., Diana, G.D., Dutko, F.J., Fancher, M., Rueckert, R.R., and Heinz, B.A. (1988). *Proc. Natl. Acad. Sci. U.S.A.* **85,** 3304–3308.
21. Rombaut, B., Brioen, P., and Boeyé, A. (1990). *J. Gen Virol.* **71,** 1081–1086.
22. Erickson, J.W., Frankenberger, E.A., Rossmann, M.G., Fout, G.S. Medappa, K.C., and Rueckert, R.R. (1983). *Proc. Natl. Acad. Sci. U.S.A.* **80,** 931–934.
23. Rossmann, M.G. (1989). *J. Biol. Chem.* **264,** 14587–14590.
24. Colonno, R., Condra, J., Mizutani, S., Callahan, P., Davies, M.-E., and Murcko, M. (1988). *Proc. Natl. Acad. Sci. U.S.A.* **85,** 5449–5453.
25. Kaplan, G., Peters, D., and Racaniello, V.R. (1990). *Science* **250,** 1596–1599.
26. Kaplan, G., Freistadt, M.S., and Racaniello, V.R. (1990). *J. Virol.* **64,** 4697–4702.
27. Minor, P.D., Ferguson, M., Evans, D.M.A., Almond, J.W., and Icenogle, J.P. (1986). *J. Gen. Virol.* **67,** 1283–1291.
28. Heinz, B.A., Rueckert, R.R., Shepard, D.A., Dutko, F.J., McKinlay, M.A., Fancher, M., Rossmann, M.G., Badger, J., and Smith, T.J. (1989). *J. Virol.* **63,** 2476–2485.
29. Roitt, I. (1991). "Essential Immunology," Blackwell Scientific Publications, London.
30. Arthos, J., Deen, K.C., Chaikin, M.A., Fornwald, J.A., Sathe, G., Sattentau, Q.J., Clapham, P.R., Weiss, R.A., McDougal, J.S., Pietropaolo, C., Axel, R., Truneh, A., Maddon, P.J., and Sweet, R.W. (1989). *Cell* **57,** 469–481.
31. Clayton, L.K., Hussey, R.E., Steinbrich, R., Ramachandran, H., Husain, Y., and Reinherz, E.L. (1988). *Nature* **335,** 363–366.
32. McClelland, A., deBear, J., Yost, S.C., Meyer, A.M., Marlor, C.W., and Greve, J.W. (1991). *Proc. Natl. Acad. Sci. U.S.A.* **88,** 7993–7997.
33. Staunton, D.E, Dustin, M.L., Erickson, H.P., and Springer, T.A. (1990). *Cell* **61,** 243–254.
34. Register, R.B., Uncapher, C.R., Naylor, A.M., Lineberger, D.W., and Colonno, R.J. (1991). *J. Virol.* **65,** 6589–6596.
35. Wang, J., Yan, Y., Garrett, T.P.J., Liu, J., Rodgers, D.W., Garlick, R.L., Tarr, G.E., Husain, Y., Reinherz, E.L., and Harrison, S.C. (1990). *Nature* **348,** 411–418.
36. Ryu, S., Kwong, P.D., Truneh, A., Porter, T.G., Arthos, J., Rosenberg, M., Dai, X., Xuong, N., Axel, R., Sweet, R.W., and Hendrickson, W.A. (1990). *Nature* **348,** 419–426.
37. Williams, A.F., and Barclay, A.N. (1988). *Annu. Rev. Immunol.* **6,** 381–405.
38. Koike, S., Ise, I., and Nomoto, A. (1991). *Proc. Natl. Acad. Sci. U.S.A.* **88,** 4104–4108.
39. Selinka, H.-C., Zibert, A., and Wimmer, E. (1991). *Proc. Natl. Acad. Sci. U.S.A.* **88,** 3598–3602.
40. Freistadt, M.F., and Racaniello, V.R. (1991). *J. Virol.* **65,** 3873–3876.

41. Morrison, M.E., and Racaniello, V.R. (1992). *J. Virol.* **66,** 2807–2813.
42. Jubelt, B., Gallez-Hawkins, B., Narayan, O., and Johnson, R.T. (1980). *J. Neuropathol. Exp. Neurol.* **39,** 138–148.
43. Jubelt, B., Narayan, O., and Johnson, R.T. (1980). *J. Neuropathol. Exp. Neurol.* **39,** 149–158.
44. Armstrong, C. (1939). *Public Health Rep.* **54,** 2302–2305.
45. La Monica, N., Meriam, C., and Racaniello, V.R. (1986). *J. Virol.* **57,** 515–525.
46. La Monica, N., Kupsky, W., and Racaniello, V.R. (1987). *Virology* **161,** 429–437.
47. Murray, M.G., Bradley, J., Yang, X.F., Wimmer, E., Moss, E.G., and Racaniello, V.R. (1988). *Science* **241,** 213–215.
48. Martin, A., Wychowski, C., Couderc, T., Crainic, R., Hogle, J., and Girard, M. (1988). *EMBO J.* **7,** 2839–2847.
49. Moss, E.G., and Racaniello, V.R. (1991). *EMBO J.* **5,** 1067–1074.
50. Yeates, T.O. Jacobson, D.H., Martin, A., Wychowski, C., Girard, M., Filman, D.J., and Hogle, J.M. (1991). *EMBO J.* **10,** 2331–2341.
51. Bodian, D., and Horstmann, D.H. (1965). *In* "Viral and Rickettsial Infections of Man" (F.L. Horsfall and I. Tamm, eds.), pp. 430–473, J.B. Lippincott, Philadelphia.
52. Kaplan, A.S. (1955). *Ann. N.Y. Acad. Sci.* **61,** 830–839.
53. Evans, C.A., Byatt, P.H., Chambers, V.C., and Smith, V.M. (1954). *J. Immunol.* **72,** 348–352.
54. Ledinko, N., Riordan, J.T., and Melnick, J.L. (1951). *Proc. Soc. Exp. Biol. Med.* **78,** 83–88.
55. Holland, J.J. (1961). *Virology* **15,** 312–326.
56. Kunin, C.M., and Jordan, W.S. 91961). *Am. J. Hyg.* **73,** 245–257.
57. Brown, R.H., Johnson, D., Ogonowski, M., and Weiner, H.L. (1987). *Ann. Neurol.* **21,** 64–70.
58. Ren, R., and Racaniello, V. (1992). *J. Virol.* **66,** 296–304.
59. Maddon, P.J., Dalgleish, A.G., McDougal, J.S., Clapham, P.R. Weiss, R.A., and Axel, R. (1986). *Cell* **47,** 333–348.
60. Shepley, M.P., Sherry, B., and Weiner, H.L. (1988). *Proc. Natl. Acad. Sci., U.S.A.* **85,** 7743–7747.
61. Ren, R., and Racaniello, V.R. (1993). *J. Infect. Dis.* **166** (in press).
62. Hurst, E.W. (1936). *Brain* **59,** 1–34.
63. Sabin, A.B. (1957). *In* "Cellular Biology, Nucleic Acids and Viruses" (T.M. Rivers, ed.), pp. 113–133, New York Academy of Science, New York.
64. Bodian, D. (1959). *In* "Viral and Rickettsial Infections of Man" (T.M. Rivers and F.L. Horsfall, eds.), pp. 479–498, J.B. Lippincott, Philadelphia.
65. Melnick, J.L. (1985). *In* "Virology" (B.N. Fields, D.M. Knipe, R.M. Chanock, J.L. Melnick, B. Roizman, and R.E. Shope, eds.), pp. 705–738, Raven Press, New York.
66. Blinzinger, K., and Anzil, A.P. (1974). *Lancet* **ii,** 1374–1375.
67. Wyatt, H.V. (1990). *Rev. Inf. Dis.* **12,** 547–556.
68. Morrison, L.A., and Fields, B.N. (1991). *J. Virol.* **65,** 2767–2772.
69. Vile, R.G., and Weiss, R.A. (1991). *Nature* **352,** 666–667.

INFLUENCES OF GROWTH FACTORS AND THEIR SIGNALING PATHWAYS IN MALIGNANCY

STUART A. AARONSON

Laboratory of Cellular and Molecular Biology, National Cancer Institute, Bethesda, Maryland

I. INTRODUCTION

GENETIC aberrations in growth factor signaling pathways are closely linked to developmental abnormalities and to a variety of chronic diseases including cancer. Malignant cells arise as a result of a stepwise progression of genetic events that include the unregulated expression of growth factors or components of their signaling pathways. I will discuss some normal aspects of growth factor signal transduction, as well as genetic aberrations in growth factor signaling pathways commonly implicated in human malignancy. Growth factors and cytokines are key components of the highly coordinated mechanisms that control the involvement of cellular interaction in normal development as well as the systemic responses of multicellular organisms to wounding and infection. There is indirect evidence that "paracrine" growth factors, which are made by one cell type but act on another, may also contribute to the clonal evolution of a tumor cell. We have, therefore, sought to identify and characterize novel paracrine-acting mitogens that play roles in the normal proliferation and differentiation of epithelial cells and may also influence progression of the most commonly occurring cancers of man. Finally, I will review the development and application of expression cDNA cloning technology to the identification and isolation of growth regulatory molecules.

II. NORMAL REGULATION OF MITOGENIC RESPONSIVENESS TO GROWTH FACTORS

Growth factors cause cells in the resting or G_0 phase to enter and proceed through the cell cycle. The mitogenic response occurs in two parts; the quiescent cell must first be advanced into the G_1 phase of the cell cycle by "competence" factors; the cell then traverses the G_1 phase, and becomes

17

The Harvey Lectures, Series 87, pages 17–34
© 1993 Wiley-Liss, Inc.

committed to DNA synthesis under the influence of ''progression'' factors (Pledger et al., 1977, 1978). Transition through the G_1 phase required sustained growth factor stimulation over a period of several hours (Fig. 1). If the signal is disrupted for a short period of time, the cell reverts to the G_0 state (Westermark and Heldin, 1985). There is also a critical period in the

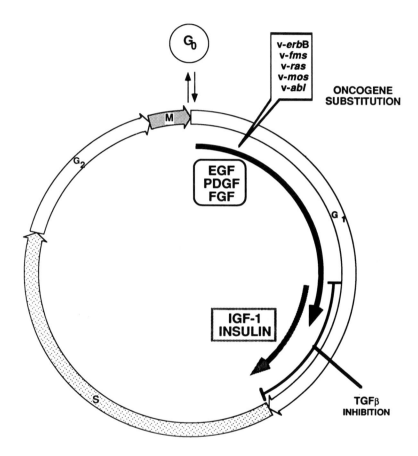

Fig. 1.   Growth factor requirements during the cell cycle. A schematic representation of requirements for the coordinated actions of two complementing growth factors to induce cell DNA synthesis. In BALB/MK cells, several oncogenes can specifically substitute for the competence factor requirement. The ability of TGFβ to inhibit the onset of DNA synthesis, even when added in late G_1, is also depicted.

G_1 phase during which simultaneous stimulation by both factors is needed to allow progression through the cell cycle (Leof et al., 1982, 1983; D. Wexler, T.P. Flemming, P.P. DiFiore, and S.A. Aaronson, unpublished observations). After this restriction point, only the presence of a "progression" factor such as insulin-like growth factor 1 (IGF-1) is needed (Pardee, 1989). Cytokines, such as transforming growth factor β (TGFβ), interferon, or tumor necrosis factor (TNF), can antagonize the proliferative effects of growth factors. In the case of TGFβ, these effects can be observed even when added relatively late in the G_1 phase (Moses et al., 1990).

In some cell types, the absence of growth factor stimulation causes the rapid onset of programmed cell death (Wyllie et al., 1987; Williams, 1991). Certain growth factors can also promote differentiation of a progenitor cell, while at the same time stimulating proliferation; others, acting on the same cell, induce only proliferation (Metcalf, 1989). Thus, there must be specific biochemical signals responsible for differentiation that only certain factors can trigger (Walker et al., 1985; Gliniak and Rohrschneider, 1990). The actions of a sequential series of growth factors can cause â hematopoietic progenitor to move through stages to a terminally differentiated phenotype (Metcalf, 1989). However, at intermediate stages, in the absence of continued stimulation by the factor, this commitment is not irreversible (Pierce et al., 1990; Rohrschneider and Metcalf, 1989). Although the differentiation program of the cell governs the diversity of phenotypic responses elicited, there are some common, highly conserved, downstream effectors of mitogenic signaling. For example, introduction of foreign receptors by DNA transfection into cells often allows coupling of the appropriate ligand to mitogenic signal transduction pathways inherently expressed by the cells (Roussel et al., 1987; Pierce et al., 1988; Van Rudden and Wagner, 1988).

III. Aberrant Growth Factor Signaling in Cancer Cells

In the early 1980s, approaches aimed at identifying the functions of retroviral oncogenes converged with efforts to investigate normal mitogenic signaling by growth factors. A number of retroviral oncogene products were found to be similar to the protein kinase encoded by v-*src* product (Collett and Erickson, 1978). Unlike many protein kinases that phosphorylate serine or threonine residues, the v-*src* product is a protein kinase that specifically phosphorylates tyrosine residues (Hunter and Sefton, 1980). Purification and sequencing of growth factors and their receptors revealed that the platelet derived growth factor (PDGF) B-chain is similar to the predicted v-*sis* oncogene product (Doolittle et al., 1983; Waterfield, 1983), and that the v-*erb*B product, which has sequence similarity to the v-*src* product, is a truncated form of the epidermal growth factor (EGF) receptor (Downward et al., 1984). Binding of EGF to its receptor results in autophosphorylation of the receptor on tyrosine (Carpenter and Cohen, 1990).

A number of other growth factors and receptors are encoded by proto-oncogenes, whose viral counterparts, like v-*sis* and v-*erb*B, were initially identified as retroviral oncogenes (Sherr et al., 1985; Besmer et al., 1986; Smith et al., 1989; Matsushime et al., 1986), or were activated by retroviral integration (Dickson et al., 1989). Others were detected as cellular oncogenes by DNA transfection (Taira et al., 1987; Delli Bovi et al., 1987; Zhan et al., 1988; Schechter et al., 1984; Dean et al., 1985; Martin-Zanca et al., 1986; Takahashi and Cooper, 1987). Still others reflect genes molecularly cloned on the basis of structural similarity to other tyrosine kinases (Aaronson, 1991; Kraus et al., 1989), or by identification of protein sequence (Ullrich et al., 1985; Yarden et al., 1986; Ullrich et al., 1986). Recently identified ligand-receptor systems include hepatocyte growth factor (HGF)-*met* (Bottaro et al., 1991; Naldini et al., 1991) and nerve growth factor (NGF)-*trk* (Hempstead et al., 1991; Nebreda et al., 1991; Klein et al., 1991), as well as two NGF-related ligands, brain-derived neurotrophic (BDNF) factor and neurotropin-3, which interact with the *trk*-B product (Sopper et al., 1991).

The PDGF system has served as the prototype for identification of substrates of the receptor tyrosine kinases. Certain enzymes become physically associated and are phosphorylated by the activated PDGF receptor kinase. These proteins include phospholipase C (PLC-γ) (Meisenhelder et al., 1989; Wahl et al., 1989), phosphatidylinositol 3' kinase (PI-3K) Kaplan et al., 1987), *ras* guanosine triphosphatase activating protein (GAP) (Molloy et al., 1989; Kaplan et al., 1990; Kazlauskas et al., 1990), and *src* and *src*-like tyrosine kinases (Ralston and Bishop, 1985; Kypta et al., 1990). These molecules contain noncatalytic domains called *src* homology (SH) regions 2 and 3. SH2 domains bind preferentially to tyrosine phosphorylated proteins and the SH3 domain may promote binding to membranes or the cytoskeleton (Koch et al., 1991). The *raf* proto-oncogene product has also been reported to become physically associated with the receptor and tyrosine phosphorylated as well (Morrison et al., 1988; Morrison et al., 1989), although the *raf* protein lacks SH2 or SH3 domains.

The connections between biochemical signals emanating from primary receptor substrates and resulting changes in the nucleus remain largely undefined. However, mitogenic signaling clearly affects the transcriptional activation of specific sets of genes, and the inactivation of others. The nuclear effectors of gene activation are transcription factors that bind to DNA as homomeric or heteromeric dimers (Mitchell and Tjian, 1989; Jones, 1990; Blackwood and Eisenman, 1991); phosphorylation appears to modulate their functions as well (Boyle et al., 1991; Gonzalez and Montminy, 1989; Ofir et al., 1990; Ayiverx and Sassone-Corsi, 1991). Within the complex regulatory network of transcription factors linked to mitogenic signaling pathways, a number, including those encoded by *jun, fos, myc, myb, rel,* and *ets,* were identified as viral oncogenes (for reviews see Eisenman, 1989; Lewin, 1991).

IV. ONCOGENE SUBVERSION OF SPECIFIC SIGNALING PATHWAYS

The evidence summarized above indicates that proto-oncogene products act at critical steps in growth factor signaling pathways. Thus, their constitutive activation as oncogene products would be expected to profoundly influence cell proliferation and possibly the differentiated state of the transformed cell. Tumor cells exhibit reduced requirements for serum in culture. The actions of oncogenes have been investigated with respect to their ability to subvert the actions of the two major growth factor signaling cascades. For instance, mouse keratinocytes can be propagated in chemically defined medium containing only two complementing growth factors, EGF and IGF-1 (Falco et al., 1988). We have shown that introduction of various oncogenically activated receptor kinases of *ras* or *raf* oncogenes completely alleviate the requirement for EGF but not IGF-1 (Fig. 1). These findings support the concept that the signaling pathway of competence factors ordinarily limits growth in vivo and that genetic changes activating critical regulatory molecules within this pathway are commonly selected during evolution of the malignant cell.

V. PARACRINE INFLUENCES ON TUMOR PROGRESSION

Growth factors released by one cell type and influencing proliferation of another cell (paracrine stimulation) may also play important roles in tumor progression. For instance, the ability of steroid hormones to stimulate epithelial cell proliferation in sex-hormone–responsive tissues such as breast and prostate appears to be mediated at least in part by hormonal effects on stromal cells (Donjacour and Cunha, 1991; Lippman and Dickson, 1989). Stromal cells, in turn, influence parenchymal cells by increasing production of growth factors, decreasing amount of inhibitory cytokines, or both. Hormonal influences on growth of breast and prostate tumors can be striking (Sukumar, 1990; Perez et al., 1989).

The chronic wounding and increased cell proliferation associated with some diseases can lead to higher risk of cancer (Coburn, 1976; Preston-Martin et al., 1990; Collins, Jr., et al., 1987). Moreover, there are a number of animal studies in which chronic injury and repair in response to agents possessing no known mutagenic actions is associated with increased cancer risk (Ames and Gold, 1990; Drinkwater, 1990). Thus, genetic lesions associated with some early cancers may allow their clonal selection in response to paracrine-acting growth factors. A case in point involves the BCL-2 oncogene, which is activated in low-grade B cell lymphomas by a chromosomal translocation of a rearranged immunoglobulin gene (Tsujimoto et al., 1984, 1985). BCL-2 acts to block apoptosis, but is not, by itself, capable of inducing cell proliferation (Vaux et al., 1988). Presumably, rescue from programmed cell death allows the BCL-2–altered cell to proliferate preferentially in response to mitogenic signals *in vivo*. This selective growth advantage can result in the eventual selection of a more malignant variant (Reed et al., 1990).

VI. NOVEL PARACRINE-ACTING EPITHELIAL CELL MITOGENS

Recognizing that most human malignancies arise in epithelial tissues where cell populations are continuously turning over, we sought to identify new growth factors for these cell types. To screen for epithelial- acting mitogens, we employed the mouse keratinocyte line BALB/MK (Weissman and Aaronson, 1983) as a prototypical epithelial cell and the NIH/3T3 fibroblast (Jainchill et al., 1969) as its nonepithelial counterpart. This approach has led to our identification and molecular characterization of two new epithelial-acting mitogens, which appear to play important roles in the normal regulation of epithelial cells in many tissues. The first, designated keratinocyte growth factor (KGF) (Rubin et al., 1989) and purified from conditioned medium of human embryonic lung fibroblasts, exhibited a unique specificity for epithelial cells (Table I). Isolation of the cDNA was accomplished by use of oligonucleotides generated on the basis of the N- terminal amino acid sequence of the purified growth factor (Finch et al., 1989). Nucleotide sequencing revealed a unique 22 kd product, although related 30% to 45% in two conserved domains to known members of the fibroblast growth factor (FGF) family.

The primary KGF translation product, like those of *hst*/K-FGF, FGF-5 and FGF- 6, contains a hydrophobic NH_2-terminal region. Evidence that this NH_2-terminal domain is not present in the mature KGF molecule indicates that it represents a signal peptide sequence (Von Heijne, 1986). Acidic and basic

TABLE I.   TARGET CELL SPECIFICITY OF KERATINOCYTE AND
HEPATOCYTE GROWTH FACTORS

| | Fold stimulation of [³H]thymidine incorporation or cell number | | | | | |
| | Epithelial cells | | | Fibroblast endothelial cells | | Melanocyte[a] |
Growth factor	B5/589	CCL208[b]	BALB/MK	NIH/3T3	HUVEC[c]	Human primary
KGF	3	10	800	<1	<1	<1
HGF	10	10	40	<1	3	180
aFGF	3	10	800	60	3	100
bFGF	3	5	200	60	3	100
EGF	15	20	200	15	ND	<1
TGFa	ND	ND	300	15	ND	<1

Comparison of maximal thymidine incorporation stimulated by HGF and other well-characterized factors in a variety of cell types, expressed as fold stimulation over untreated cultures. Results for endothelial cells were from proliferation assays, where increased cell number was measured at 7 days. These data are representative of several experiments.

ND, not determined.

[a]Primary cultures of human melanocytes were prepared and tested.
[b]A rhesus monkey bronchial epithelial line (Rubin et al., 1989). Growth factor activity was also demonstrated on primary cultures of normal human bronchial epithelial cells.
[c]Human umbilical vein endothelial cells.

FGF are apparently synthesized without signal peptides (Jaye et al., 1986; Abraham et al., 1986). The *int*-2–encoded protein contains an atypically short region of NH_2-terminal hydrophobic residues (Moore et al., 1986), which apparently functions as a signal sequence (Acland et al., 1990). The *int*-2– and FGF- 5–encoded proteins also contain COOH-terminal extensions which are longer than that of the other family members.

To investigate the functional role of KGF, we examined the expression of its transcript in a variety of human cell lines and tissues. The predominant 2.4 kb KGF transcript was detected in each of several stromal fibroblast lines derived from epithelial tissues of embryonic, neonatal, and adult sources. In contrast, the transcript was not detected in normal glial cells, nor in a variety of epithelial cell lines. The transcript was also evident in RNA extracted from normal adult kidney and organs of the GI tract, but not from lung or brain (Finch et al., 1989). The striking specificity of KGF RNA expression in stromal cells from epithelial tissues supports the concept that this factor is important in the normal mesenchymal stimulation of epithelial cell growth.

We purified a second growth factor from conditioned medium of the same embryonic fibroblast line. This molecule, like KGF, exhibited a novel target cell specificity including epithelial cells, melanocytes, and endothelial cells (Table I). N-terminal sequence analysis of the purified growth factor revealed a unique amino acid sequence as well. In the course of our attempts to molecularly clone this growth factor, reports describing the cloning of a molecule, designated hepatocyte growth factor (HGF) appeared (Miyazawa et al., 1989; Nakamura et al., 1989). Cloning and characterization of our fibroblast-derived mitogen revealed its identity with HGF (Rubin et al., 1991).

HGF was initially detected as a hormonelike activity capable of stimulating hepatocyte proliferation (Nakamura et al., 1987; Gohda et al., 1988; Zarnegar and Michalopoullous, 1989). This growth factor is expressed by stromal fibroblasts and is mitogenic for a variety of cell types (Rubin et al., 1991, Kan et al., 1991; Igawa et al., 1991; Matsumoto et al., 1991). HGF is also highly related or identical to scatter factor, an agent that stimulates the dispersion of epithelial and vascular endothelial cells (Gherardi and Stoker, 1990; Weidner et al., 1990; Rosen et al., 1990). Structurally, HGF resembles plasminogen, in that HGF has characterisic NH_2-terminal kringle domains (Patthy et al., 1984) and a COOH-terminal serine protease-like domain (Miyazawa et al., 1989; Nakamura et al., 1989). Each kringle is an approximately 80-amino acid stretch containing a characteristic set of three internal disulfide bonds. HGF is synthesized as an 87 kd single chain polypeptide (p87). Like plasminogen, it can be cleaved into a heterodimeric form consisting of a heavy (~60 kd) and a light chain (~30 kd) held together by disulfide bonds (Nakamura et al., 1987; Gohda et al., 1988; Zarnegar and Michalopoulous, 1989).

We have observed expression of HGF in stromal fibroblasts derived from adult kidney, lung, gastrointestinal tract, and prostate, as well as in embryonic lung fibroblasts (unpublished observations). Others have reported detection of an HGF transcript (Tashiro et al., 1990) or immunologically cross-reactive material (Zarnegar et al., 1990) in several organs. Moreover, expression of the HGF transcript in liver was demonstrated in the nonparenchymal component as opposed to hepatocytes (Kinoshita et al., 1989). Thus, this factor appears to be widely expressed and likely to act in a paracrine fashion on a wide variety of cell types.

We also purified a naturally occurring HGF variant, whose predicted sequence extends only through the second kringle domain of this plasminogen-related molecule (Fig. 2). This smaller molecule, derived from an alternative HGF transcript, lacked mitogenic activity but specifically inhibited HGF-induced mitogenesis. Cross-linking studies demonstrated that the trun-

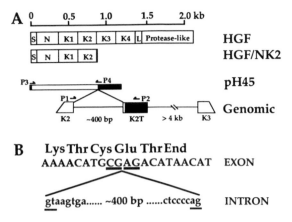

Fig. 2. Characterization of an HGF/NK2 cDNA. **A:** Schematic representation of the domain structures of HGF and HGF/NK2 (open boxes). The 1.2 kb cDNA clone pH45 is composed of a coding region (open bar) and untranslated ones (filled bars). The splicing event for the generation of the HGF/NK2 sequence is depicted by the genomic region, with the alternative exon (K2T) located ~400 bp downstream of the K2 exon. The intronic region between K2T and the downstream K3 exon is of undefined length. Arrows represent the positions and directions of the polymerase chain reaction (PCR) primers used. Abbreviations are as follows: K1 to K4, kringle 1 to 4; and L, linker region. Sizes in kilobases are shown at the top. The double lines at the end of HGF/NK2 indicate the two additional amino acids. **B:** (Top) The cDNA and the predicted amino acid sequences of HGF/NK2 (exon) at the point of divergence with HGF are shown with the splice site underlined. (Bottom) The corresponding genomic region (intron) includes a ~400 bp intron with the consensus splicing signals at the exon-intron boundaries underlined.

cated molecule competes with HFG for binding to the c-*met* proto-oncogene product. Thus, the same gene encodes both a growth factor and its direct antagonist (Chan et al., 1991).

Secreted forms of certain growth factor receptors containing only their external domains may act to transport or protect the respective ligand (Ullrich et al., 1984; DiStefano and Johnson, 1988; Leung et al., 1987; Johnson et al., 1990; Mosley et al., 1989). In contrast, HGF/NK2 competes directly with HGF for binding to a common cell surface receptor. The interleukin-1 (IL-1) antagonist IRAP is a nonmitogenic, competitive inhibitor of IL-1 (Hannum et al., 1990; Eisenberg et al., 1990; Carter et al., 1990). However, IRAP and IL-1 share only approximately 25% amino acid sequence identity and are encoded by distinct genes. By contrast, HGF and HGF/NK2 protein sequences are more than 99% identical throughout the entire length of the smaller molecule and are encoded as alternative transcripts of the same gene.

We observed considerable variation in relative concentrations of HGF and HGF/NK2 transcripts among the fibroblasts analyzed, which suggests that an HGF/NK2 excess could exist under some physiological settings. The HGF/NK2 transcript appears to be more abundant in quiescent cells (unpublished data). This would be consistent with a model in which HGF/NK2 modulates HGF growth-stimulating effects at late stages of wound healing and tissue repair as fibroblasts return to the resting state. Conceivably, HGF/NK2 could be used to inhibit the growth of tumors in which HGF was implicated.

The identification of factors such as KGF and HGF may help to explain how stromal cells indirectly influence normal epithelial cell growth in response to hormones and how these cells provide an environment conducive to proliferation, invasion, and even metastasis of epithelial tumor cells. Another aspect of malignancy is neo-angiogenesis, the process by which new vasculature is developed to support nourishment of malignant cells (Liotta et al., 1991). A number of growth factors including the EGFs, FGF, and HGF are chemotatic for endothelial cells and induce their proliferation. Such angiogenic factors may be released by stromal cells as part of an aberrant wound-healing response to tumor cell proliferation, or by the tumor cell itself. In such a manner, they may contribute indirectly to tumor progression.

VII. EXPRESSION cDNA CLONING OF GROWTH REGULATORY MOLECULES

We have recently reported the development of an efficient system for construction of unidirectional cDNA libraries (Miki et al., 1989), which has made it possible to isolate full-length coding sequences of several genes (Finch et al., 1989; Kraus et al., 1989; Kruh et al., 1990; Matsui et al., 1989; Miki et al., 1989; Rubin et al., 1991). The lambda phage plasmid cDNA expression vector (λpCEV)27 system was developed to clone cDNAs by means of stable phenotypic changes induced in the recipient cell (Miki et al., 1991b). Use of a λ-plasmid composite vector made it possible to generate high-complexity cDNA libraries and to efficiently excise the plasmid from the stably integrated phagemid DNA. This phagemid vector (Fig. 3) contained several features that allow for construction of cDNA libraries using the automatic directional cloning (ADC) method (Miki et al., 1989) (Fig. 3).

The capacity for efficient rescue of cDNA clones from mammalian cells

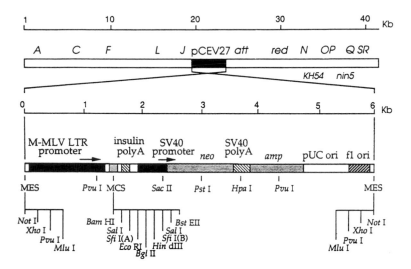

Fig. 3. Structure of the λpCEV27 cDNA expression cloning vector. Structure of the λpCEV27 vecetor is shown in the upper half with the location of λ genes. KH54 and *nin*5 are the deletions in the λ genome. The plasmid portion is enlarged and shown in the lower half. The vector contains several features including two *Sfi*I sites in multiple cloning sites (MCS) for construction of cDNA libraries using the automatic directional cloning (ADC) method, a Moloney murine leukemia virus long terminal repeat (M-MLV LTR) promoter suitable for cDNA expression in mammalian cells, the rat preproinsulin poly-A signal downstream from the cDNA cloning site, the SV40 promoter-drive *neo* gene as a selectable marker, and multiple excision sites (MESs) for plasmid rescue from genomic DNA. A stuffer fragment encoding a part o the *lac*Z gene (*lac*Z′) is present in the *Bgl*II site in MCS. The *f*1 replication origin, as well as SP6 and T7 phage promoters are also included to facilitate construction of subtraction libraries, as well as analysis of cDNA inserts, by production of single-stranded DNA and RNA transcripts, respectively.

is another important feature of this stable expression cloning system. When plasmid cDNA libraries are used to transfect mammalian cells, single plasmids integrated in genomic DNA are difficult to release. Plasmid rescue is readily achieved only when multiple copies are clustered at a single integration site. In contrast, when the λ-plasmid composite vector is used for cDNA library construction, cDNA clones are integrated with recombination between λ DNA and host genomic DNA. This generally leaves intact the plasmid nestled within the λ arms. For plasmid rescue, genomic DNA extracted from the transformant is subjected to digestion with an enzyme which can cleave the λ-plasmid junctions. The resulting DNA can then be circularized and used for bacterial transformation.

We tested the efficacy of our expression cDNA cloning system in an effort to isolate the receptor for KGF. To do so, we took advantage of the fact that NIH/3T3 fibroblasts lack specific binding sites for this growth factor, are susceptible to uptake and stable integration of transfected DNA molecules, and endogenously express KGF. The availability of the highly efficient cDNA expression vector system made it possible to test whether transfection with a cDNA library prepared from an epithelial cell that expressed KGF receptors might activate an autocrine transforming loop in NIH/3T3 cells. If so, it would allow recognition of the transformed phenotype of the recombinant cell expressing KGF receptors.

We prepared a cDNA library (4.5×10^6 independent clones) from BALB/MK epidermal keratinocytes in λpCEV27. Transfection of NIH/3T3 mouse embryo fibroblasts which synthesize KGF, by the library DNA, led to detection of 15 transformed foci among a total of 100 individual cultures. Each focus was tested and shown to be resistant to geneticin, indicating that it contained integrated vector sequences. Three representative transformants were chosen for more detailed characterization based upon differences in their morphologies.

When we performed plasmid rescue, each transformant gave rise to at least three distinct cDNA clones as determined by physical mapping. To examine their biological activities, each clone was subjected to transfection analysis on NIH/3T3 cells. A single clone rescued from each transformant was found to possess high-titer transforming activity ranging from 10^3 to 10^4 focus-forming units/per nanomole of DNA. Moreover, the morphology of the foci induced by each cDNA was similar to that of the parental transformant. Because of their distinct physical maps and distinguishable biological properties, we designated the genes for these transforming cDNAs as *ect*1, *ect*2, and *ect*3 (for *e*pithelial *c*ell *t*ransforming genes). Transfectants induced by the individual transforming plasmids were utilized in subsequent analyses.

To further investigate the possibility that any of the transfectants might encode the KGF receptor, we performed binding studies with recombinant [^{125}I]-KGF as the tracer molecule. BALB/MK cells demonstrated specific high-affinity binding of [^{125}I]-KGF, while there was no such binding detectable to NIH/3T3 cells. Of note, expression of the *ect*1 gene by NIH/3T3 cells resulted in the acquisition of 3.5 times as many [^{125}I]-KGF binding sites as expressed by BALB/MK cells (Miki et al., 1991a). Under these same conditions, neither NIH/*ect*2 nor NIH/*ect*3 bound significant levels of the labeled growth factor. These results strongly suggested that *ect*1 encoded the KGF receptor, whose introduction into NIH/3T3 cells had completed an autocrine transforming loop.

Characterization of the cloned 4.2 kb cDNA revealed a predicted membrane-spanning tyrosine kinase structurally related to the FGF receptor (FGFR) (Miki et al., 1991a). As shown in Figure 4, structural analysis of the human KGF receptor cloned by the analogous procedure revealed identity with one of the fibroblast growth factor (FGF) receptors (*bek*/FGFR-2) except for a divergent stretch of 49 amino acids in their extracellular domains (Miki et al., 1992). Binding assays demonstrated that the KGF receptor was a high-affinity receptor for both KGF and acidic FGF, while FGFR-2 showed high affinity for basic FGF and acidic FGF but no detectable binding by KGF (Fig. 4). Analysis of the *bek* gene revealed two alternative exons responsible for the region of divergence between the two receptors. The KGF receptor (R) transcript was specific to epithelial cells, and it appeared to be differentially regulated with respect to the alternative FGFR-2 transcript (Miki et al., 1992). The strikingly different ligand-binding affinities of these two receptors encoded by a single gene, combined with their different patterns of expression, open a new dimension to growth factor receptor diversity and may reflect a general mechanism for increasing the repertoire of these important cell surface molecules.

We isolated at least two additional genes from our epithelial cell cDNA expression library that showed transforming activity for fibroblasts. Their relationships to growth-signaling pathways awaits further characterization. Given the normal juxtaposition of other cell types and stromal fibroblasts *in vivo* and their likely paracrine interactions, it may be possible to identify additional growth factors or their receptors by transfecting NIH/3T3 cells with expression libraries from these other sources. It is conceivable that introduction of tissue-specific genes that encode downstream targets of growth factor receptor signaling might complete an autocrine loop in a recipient cell, if it played a critical regulatory role. Thus, our expression cloning strategy may aid in the identification and isolation of genes for such intracellular components of mitogenic signaling pathways. To detect an oncogene by gene transfer, it must be small enough to be transfected, and its promoter must allow a high level of expression in the recipient cell. These problems can be overcome by use of efficient cloning vectors allowing stable expression. It seems likely that, as these approaches are combined with efforts to increase the efficiency of stable transfection of recipient cells other than NIH/3T3, it may also be possible to identify and isolate novel oncogenes of cancer cells.

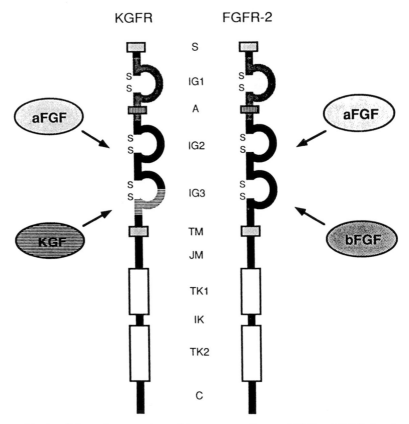

Fig. 4.   Schematic comparison of the structure of human KGFR and FGFR-2 and domains responsible for binding of acidic FGF, basic FGF, and KGF.

ACKNOWLEDGMENTS

I gratefully acknowledge many dedicated coworkers who have contributed importantly to our research. I am particularly indebted to Jeff Rubin, Paul Finch, Andrew Chan, Don Bottaro, Tim Fleming, and Toru Miki for their contributions to this work. I am also fortunate to have had the continuing support of the National Cancer Institute in our research efforts.

REFERENCES

Aaronson, S.A. (1991). *Science* **254,** 1146.
Abraham, J.A., Mergia, A., Whang, J.L., Tumolo, A., Friedman, J., Hjerrild, K.A., Gospodarowicz, D., and Fiddes, J.C. (1986). *Science* **233,** 545.

Acland, P., Dixon, M., Peters, G., and Dickson, C. (1990). *Nature* **343,** 662.

Ames, B.N., and Gold, L.S. (1990). *Science* **249,** 970.

Ayiverx, J., and Sassone-Corsi, P. (1991). Cell **64,** 983.

Besmer, P., Murphy, J.E., George, P.C., Qiu, F.H., Bergold, P.J., Lederman, L., Snyder, H.W., Jr., Brodeur, D., Zuckerman, E.E., and Hardy, W.D. (1986). *Nature* **320,** 415.

Blackwood, E.M., and Eisenman, R.N. (1991). *Science* **251,** 1211.

Bottaro, D.P., Rubin, J.S., Faletto, D.L., Chan, A.M.-L., Kmiecik, T.E., Vande Woude, G.F., and Aaronson, S.A. (1991). *Science* **251,** 802

Boyle, N.J., Smeal, T., Defize, L.H.K., Angel, P., Woodgett, J.R., Karin, M., and Hunter, T. (1991). *Cell* **64,** 573.

Carpenter, G., and Cohen, S. (1990). *J. Biol. Chem.* **165,** 7709.

Carter, D.B., Deibel, M.R., Jr., Dunn, C.J., Tomich, C.S., Laborde, A.L., Slightom, J.L., Berger, A.E., Bienkowski, M.J., Sun, F.F., and McEwan, R.N. (1990). *Nature,* **344,** 633.

Chan, A. M.-L., Rubin, J.S., Bottaro, D.P., Hirschfield, D.W., Chedid, M., and Aaronson, S.A. (1991). *Science* **254,** 1382.

Coburn, R.J. (1976). ''Cancer of the Skin,'' pp. 939–949, W.B. Saunders, Philadelphia.

Collett, M., and Erickson, R. (1978). *Proc. Natl. Acad. Sci. U.S.A.* **75,** 2021.

Collins, R.H., Jr., Feldman, M., and Fordtran, J.S. (1987). *N. Engl. J. Med.* **316,** 1654.

Dean, M., Park, M., Le Beau, M.M., Robins, T.S., Diaz, M.O., Rowley, J.D., Blair, D.G., and Vande Woude, G.F. (1985). *Nature* **318,** 385.

Delli Bovi, P., Curatola, A.M., Kern, F.G., Greco, A., Ittmann, M., and Basilico, C. (1987). *Cell* **50,** 729.

Dickson, C., Deed, R., Dixon, M., and Peters, G. (1989). *Prog. Growth Fact. Res.* **1,** 123.

DiStefano, P.S., and Johnson, Jr. E.M., (1988). *Proc. Natl. Acad. Sci. U.S.A.* **85,** 270.

Donjacour, A.A., and Cunha, G.R. (1991). *Cancer Treat. Res.* **53,** 335.

Doolittle, R., Hunkapiller, M.W., Hood, L.E., Devare, S.G., Robbins, K.C., Aaronson, S.A., and Antoniades, H.N. (1983). *Science* **221,** 275.

Downward, J. (1984). *Nature* **307,** 521.

Drinkwater, N.R. (1990). *Cancer Cells* **2,** 8.

Eisenberg, S.P., Evans, R.J., Arend, W.P., Verderber, E., Brewer, M.T., Hannum, C.H., and Thompson, R.C. (1990). *Nature* **343,** 341.

Eisenman, R.N. (1989). *In* ''Oncogenes and the Molecular Origin of Cancer'' (R.E. Weinberg, ed.), pp. 175–221, Cold Spring Harbor Laboratory, Cold Spring Harbor, NY.

Falco, J.P., Taylor, W.G., Di Fiore, P.P., Weissman, B.E., and Aaronson, S.A. (1988). *Oncogene* **2,** 573.

Finch, P.W., Rubin, J.S., Miki, T., Ron, D., and Aaronson, S.A. (1989). *Science* **245,** 752.

Gherardi, E., and Stoker, M. (1990). *Nature* **346,** 228.

Gliniak, G.C., and Rohrschneider, L.R. (1990). *Cell* **63,** 1073.

Gohda, E., Tsubouchi, H., Nakayama, H., Hironi, S., Sakiyama, O., Takahashi, K., Miyazaki, H., Hashimoto, S., and Daikuhara, Y. (1988). *J. Clin. Invest.* **81,** 414.

Gonzalez, C.A., and Montminy, M.R. (1989). *Cell* **59,** 675.

Hannum, C.M., Wilcox, C.J., Arend, W.P., Joslin, F.G., Dripps, D.J., Heimdal, P.L., Armes, L.G., Sommer, A., Eisenberg, S.P., and Thompson, R.C. (1990). *Nature,* **343,** 336.

Hempstead, B.L., Martin-Zanca, D., Kaplan, D.R., Parada, L.F., and Chao, M.V. (1991). *Nature* **350,** 678.

Hunter, T., and Sefton, B. (1980). *Proc. Natl. Acad. Sci. U.S.A.* **77,** 1311.

Igawa, T., Kanda, S., Kanetake, H., Saitoh, Y., Ichihara, A., Tomita, Y., and Nakamura, T. (1991). *Biochem and Biophys. Res. Commun.* **174,** 831.

Jainchill, J.S., Aaronson, S.A., and Todaro, G.J. (1969). *J. Virol.* **4,** 549.

Jaye, M., Howk, R., Burgess, W., Ricca, G.A., Chiu, I.M., Ravera, M.W., O'Brien, S.J., Modi, W.S., Maciag, T., and Drohan, W.N. (1986). *Science* **233,** 541.

Jones, N. (1990). *Cell* **61,** 9.

Johnson, D.E., Lee, P.L., Lu, J., and Williams, L.T. (1990). *Mol. and Cell. Biol.* **10,** 4728.

Kan, M., Zhang, G.H., Zarnegar, R., Michalopoulos, G., Myoken, Y., McKeehan, W.L., and Stevens, J.I. (1991). *Biochem. Biophys. Res. Commun.* **174,** 331.

Kaplan, D.R., Morrison, D.K., Wong, G., McCormick, F., and Williams, L.T. (1990). *Cell* **61,** 125.

Kaplan, D.R., Whitman, M., Schaffhausen, B., Pallas, D.C., White, M., Cantley, L., and Roberts, T.M. (1987). *Cell* **50,** 1021.

Kazlauskas, A., Ellis, C., Pawson, T., and Cooper, J.A. (1990). *Science* **247,** 1578.

Kinoshita, T., Tashiro, K., and Nakamura, T. (1989). *Biochem. Biophys. Res. Commun.* **165,** 1229.

Klein, R., Jing, S.Q., Nanduri, V., O'Rourke, E., and Barbacid, M. (1991). *Cell* **65,** 189.

Koch, C.A., Anderson, D., Moran, M.F., Ellis, C., and Pawson, T. (1991). *Science* **252,** 668.

Kraus, M.H., Issing, W., Miki, T., Popescu, N.C., and Aaronson, S.A. (1989). *Proc. Natl. Acad. Sci. U.S.A.* **86,** 9193.

Kruh, G.D., Perego, R., Miki, T., and Aaronson, S.A. (1990). *Proc. Natl. Acad. Sci. U.S.A.* **87,** 5802.

Kypta, R.M., Goldberg, Y., Ulug, E.T., and Courtneidge, S.A. (1990). *Cell* **62,** 481.

Leof, E.B., Van Wyk, J.J., O'Keefe, E.J., and Pledger, W.J. (1983). *Exp. Cell Res.* **147,** 202.

Leof, E.B., Wharton, W., Van Wyk, J.J., and Pledger, W.J. (1982). **141,** 107.

Leung, D.W., Spencer, S.A., Cachianes, G., Hammonds, R.G., Collins, C., Henzel, W.J., Barnard, R., Waters, M.J., and Wood, W.I. (1987). *Nature* **330,** 537.

Lewin, B. (1991) *Cell* **64,** 303.

Liotta, L., Steeg, P.S., and Stetler-Stevenson, W.G. (1991). *Cell* **64,** 327.

Lippman, M.E., and Dickson, R.B. (1989). *J. Steroid Biochem.* **34,** 107.

Martin-Zanca, D., Hughes, S.H., Barbacid, M. (1986). *Nature* **319,** 743.

Matsui, T., Heidaran, M., Miki, T., Popescu, N., La Rochelle, W., Kraus, M., Pierce, J., and Aaronson, S.A. (1989). *Science* **243,** 800.

Matsumoto, K., Tajima, H., and Nakamura, T. (1991). *Biochem. Biophys. Res. Commun.* **170,** 45.

Matsushime, H., Wang, L.H., and Shibuya, M. (1986). *Mol. Cell. Biol.* **6,** 3000.

Meisenhelder, J., Suh, P.-G., Rhee, S.G., and Hunter, T. (1989). *Cell* **57,** 1109.

Metcalf, D. (1989). *Nature* **339,** 27.

Miki, T., Matsui, T., Heidaran, M.A., and Aaronson, S.A. (1989). *Gene* **83,** 137.

Miki, T., Fleming, T.P., Bottaro, D.P., Rubin, J.S., Ron, D., and Aaronson, S.A. (1991a). *Science* **251,** 72.

Miki, T., Fleming, T.P., Crescenzi, M., Molloy, C.J., Blam, S.B., Reynolds, S.H., and Aaronson, S.A. (1991b). *Proc. Natl. Acad. Sci. U.S.A.* **88,** 5167.

Miki, T., Bottaro, D.P., Fleming, T.P., Smith, C.L., Burgess, W.H., Chan, A.M.-L., and Aaronson, S.A. (1992). *Proc. Natl. Acad. Sci. U.S.A.* **89,** 246.

Mitchell, P.J., and Tjian, R. (1989). *Science* **245,** 371.

Miyazawa, K., Tsubouchi, H., Naka, D., Takahashi, K., Okigaki, M., Arakaki, N.,

Nakayama, H., Hirono, S., Sakiyama, O., and Takahashi, K. (1989). *Biochem. Biophys. Res. Commun.* **163,** 967.

Molloy, C.J., Bottaro, D.P., Fleming, T.P., Marshall, M.S., Gibbs, J.B., and Aaronson, S.A. (1989). *Nature* **342,** 711.

Moore, R., Casey, G., Brookes, S., Dixon, M., Peters, G., and Dickson, C. (1986). *EMBO J.* **5,** 919.

Morrison, D.K., Kaplan, D.R., Rapp, U., and Roberts, T.M. (1988). *Proc. Natl. Acad. Sci. U.S.A.* **85,** 8855.

Morrison, D.K., Kaplan, D.R., Escobedo, J.A., Rapp, U.R., Roberts, T.M., and Williams, L.T. (1989). *Cell* **581,** 649.

Moses, H.L., Yang, E.Y., and Pietenpol, J.A. (1990). *Cell* **63,** 245.

Mosley, B., Beckmann, M.P., March, C.J., Idzerda, R.L., Gimpel, S.D., VandenBos, T., Friend, D., Alpert, A., Anderson, D., and Jackson, J. (1989). *Cell* **59,** 335.

Nakamura, T., Nawa, K., Ichihara, A., Kaise, N., and Nishino, T. (1987). *FEBS Lett.* **224,** 311.

Nakamura, T., Nishizawa, T., Hagiya, M., Seki, T., Shimonishi, M., Sugimura, A., Tashiro, K., and Shimizu, S. (1989). *Nature* **342,** 440.

Naldini, L., Vigna, E., Narshimhan, R.P., Gaudino, G., Zarnegar, R., Michalopoulos, G.K., and Comoglio, P.M. (1991). *Oncogene* **6,** 501.

Nebreda, A.R., Martin-Zanca, D., Kaplan, D.R., Parada, L.F., and Santos, E. (1991). *Science* **252,** 558.

Ofir, R., Dwarki, V.G., Rashid, D., and Verma, I.M. (1990). *Nature* **348,** 78.

Pardee, A.B. (1989). *Science* **246,** 603.

Patthy, L., Trexler, M., Vali, Z., Banyai, L., and Varadi, A. (1984). *FEBS Lett* **171,** 131.

Perez, C.A., Fair, W.R., and Ihde, D.C. (1989). *In* "Cancer: Principles and Practice of Oncology" (V.T. De Vita, S.R. Hallman, and S.A. Rosenberg, eds.), pp. 1023–1058, J.B. Lippincott, Philadelphia.

Pierce, J.H., Ruggiero, M., Fleming, T.P., Di Fiore, P.P., Greenberger, J.S., Varticovski, L., Schlessinger, J., Rovera, G., and Aaronson, S.A. (1988). *Science* **239,** 628.

Pierce, J.H., Di Marco, E., Cox, G.W., Lombardi, D., Ruggiero, M., Varesio, L., Wang, L.M., Choudhury, G.G., Sakaguchi, A.Y., and Di Fiore, P.P. (1990). *Proc. Natl. Acad. Sci. U.S.A.* **87,** 5613.

Pledger, W.J., Stiles, C.D., Antoniades, H.N., and Scher, C.D. (1978). *Proc. Natl. Acad. Sci. U.S.A.* **75,** 2839.

Pledger, W.J., Stiles, C.D., Antoniades, H.N., and Scher, C.D. (1977). *Proc. Natl. Acad Sci. U.S.A.* **74,** 4481.

Preston-Martin, S., Pike, M.C., Ross, R.K., Jones, P.A., and Henderson, B.E. (1990). *Cancer Res.* **50,** 7415.

Ralston, R., and Bishop, J.M. (1985). *Proc. Natl. Acad. Sci. U.S.A.* **82,** 7845.

Reed, J.C., Cuddy, M., Haldar, S., Croce, C., Nowell, P., Makover, D., and Bradley, K. (1990). *Proc. Natl. Acad. Sci. U.S.A.* **87,** 3660.

Rohrschneider, L., and Metcalf, D. (1989). *Mol. Cell. Biol.* **9,** 5081.

Rosen, E., Meromsky, L., Setter, E., Vinter, D.W., and Goldberg, I.D. (1990). *Exp. Cell Res.* **186,** 22.

Roussel, M.F., Dull, T.J., Rettenmier, C.W., Ralph, P., Ullrich, A., and Sherr, C.J. (1987). *Nature* **325,** 549.

Rubin, J.S., Osada, H., Finch, P.W., Taylor, W.G., Rudikoff, S., and Aaronson, S.A. (1989). *Proc. Natl. Acad. Sci. U.S.A.* **86,** 802.

Rubin, J.S., Chan, A.M.-L., Bottaro, D.P., Burgess, W.H., Taylor, W.G., Cech, A.C.,

Hirschfield, D.W., Wong, J., Miki, T., and Finch, P.W. (1991). *Proc. Natl. Acad. Sci. U.S.A.* **88,** 415.

Schechter, A.L., Stern, D.F., Vaidyanathan, L., Decker, S.J., Drebin, J.A., Greene, M.I., and Weinberg, R.A. (1984). *Nature* **312,** 513.

Sherr, C.J., Rettenmier, C.W., Sacca, R., Roussel, M.F., Look, A.T., and Stanley, E.R. (1985). *Cell* **41,** 665.

Smith, D.R., Vogt, P.K., and Hayman, M.J. (1989) *Proc. Natl. Acad. Sci. U.S.A.* **86,** 5291.

Soppet, D., Escandon, E., Maragos, J., Reed, S.W., Blair, J., Burton, L.E., Stanton, B.R., Kaplan, D.R., Hunter, T., Nikolics, K., and Parada, L.F. (1991). *Cell* **65,** 895.

Sukumar, S. (1990). *Cancer Cells* **2,** 199.

Taira, M., Yoshida, T., Miyagawa, K., Sakamoto, H., Terada, M., and Sugimura, T. (1987). *Proc. Natl. Acad. Sci. U.S.A.* **84,** 2980.

Takahashi, M., and Cooper, G.M. (1987). *Mol. Cell. Biol.* **7,** 1378.

Tashiro, K., Hagiya, M., Nishizawa, T., Seki, T., Shimonishi, M., Shimizu, S., and Nakamura, T. (1990). *Proc. Natl. Acad. Sci. U.S.A.* **87,** 3200.

Tsujimoto, Y., Finger, L.R., Yunis, J., Nowell, P.C., and Croce, C.M. (1984). *Science* **226,** 1097.

Tsujimoto, Y., Cossman, J., Jaffe, E., and Croce, C.M. (1985). *Science* **228,** 1440.

Ullrich, A., Coussens, L., Hayflick, J.S., Dull, T.J., Gray, A., Tam, A.W., Lee, J., Yarden, Y., Libermann, T.A., and Schlessinger, J. (1984). *Nature* **309,** 418.

Ullrich, A., Bell, J.R., Chen, E.Y., Herrera, R., Petruzzelli, L.M., Dull, T.J., Gray, A., Coussens, L., Liao, Y.C., and Tsubokawa, M. (1985). *Nature* **313,** 756.

Ullrich, A., Gray, A., Tam, A.W., Yang-Gend, T., Tsubokawa, M., Collins, C., Henzel, W., Le Bon, T., Kathuria, S., and Chen, E. (1986). *EMBO J.* **5,** 2503.

Van Rudden, J., and Wagner, I. (1988). *EMBO J.* **7,** 2749.

Vaux, D.L., Cory, S., and Adams, J.M. (1988). *Nature* **335,** 440.

Von Heijne, G. (1986). *Nucleic Acids Res.* **14,** 4683.

Wahl, M.I., Olashaw, N.E., Nishibe, S., Rhee, S.G., Pledger, W.J., and Carpenter, G. (1989). *Mol. Cell. Biol.* **9,** 2934.

Walker, F., Nicola, N.A., Metcalf, D., and Burgess, A.N. (1985). *Cell* **43,** 269.

Waterfield, M. (1983). *Nature* **304,** 35.

Weidner, K.M., Behrens, J., Vanderkerckhove, J., and Birchmeier, W. (1990). *J. Cell Biol.* **111,** 2097.

Weissman, B.E., and Aaronson, S.A. (1983). *Cell* **32,** 599.

Westermark, B., and Heldin, C.H. (1985). *J. Cell Physiol.* **124,** 43.

Williams, G.T. (1991). *Cell* **65,** 1097.

Wyllie, A.H., Rose, K.A., Morris, R.G., Steel, C.M., Foster, E., and Spandidos, D.A. (1987). *Br. J. Cancer* **56,** 251.

Yarden, Y., Escobedo, J.A., Kuang, W.J., Yang-Feng, T.L., Daniel, T.O., Tremble, P.M., Chen, E.Y., Ando, M.E., Harkins, R.N., and Francke, U. (1986). *Nature* **323,** 226.

Zarnegar, R., and Michalopoulous, G. (1989). *Cancer Res.* **49,** 3314.

Zarnegar, R., Muga, S., Rahua, R., and Michalopoulos, G. (1990). *Proc. Natl. Acad. Sci. U.S.A.* **87,** 1252.

Zhan, X., Bates, B., Hu, X.G., and Goldfarb, M. (1988). *Mol. Cell Biol.* **8,** 3487.

THE STRUCTURE AND FUNCTION OF DNA SUPERCOILING AND CATENANES

NICHOLAS R. COZZARELLI

Department of Molecular and Cell Biology, University of California, Berkeley, California

WHEN I was a graduate student in the Department of Biochemistry at Harvard, the required course reading often consisted of very long, very dry review articles. As such, I used to look forward to reading the Harvey Lectures for these courses, as much for what they said about scientists as what they said about science. I am honored and humbled now to be a producer, as well as a consumer, of Harvey Lectures. I will present the results of two recent projects—the structure of supercoiled DNA and the topology of kinetoplast networks. In the spirit of what was valuable to me as a student, I will try to give more than usual of the thinking behind the experiments, and less of the data.

I hope to convince the reader that we now know a great deal about the conformation of supercoiled DNA in solution and have a fascinating start on determining the topology of the kinetoplast DNA network. I also want to convince you of something else. Everything in this article required a collaboration with mathematicians, and I believe that mathematics will make an increasingly important contribution to biology. Such interdisciplinary collaborations are becoming increasingly easy to effect. There are many mathematicians who look forward to making a biologically relevant contribution, and many biological problems would benefit from a quantitative approach. My advice is to try it; I think you might like it.

The DNA in all organisms is negatively supercoiled, and to about the same degree (Bauer, 1978). DNA supercoiling is best considered in terms of the concept of the linking number (Lk), which is a measure of the crossing of the two strands of the DNA double helix (Cozzarelli et al., 1990). The linking number of relaxed DNA is designated Lk_0. The statement, then, that DNA is negatively supercoiled means that ΔLk, $Lk - Lk_0$, is a negative number for

35

The Harvey Lectures, Series 87, pages 35–55
© 1993 Wiley-Liss, Inc.

all naturally occurring DNA. It is often more convenient to express DNA supercoiling by the length-independent quantity obtained by dividing ΔLk by Lk_0, which is the supertwist density, or σ. The value of σ in all organisms, prokaryotic and eukaryotic, is about -0.06 (Bauer, 1978). This means that for about every 16 turns of the strands of the DNA double helix about each other, there is one opposite crossing provided by supercoiling. Because linear DNA in vivo is divided into topologically constrained domains, it too is negatively supercoiled. Lk is a topological invariant, i.e., a property of DNA that can only be changed by breakage and reunion of its backbone. Linking number can have only integral values. Remarkably, Lk is the sum of two continuously varying geometric properties, twist and writhe. Twist is a measure of the winding of the two strands of DNA about each other that is imparted by its double-helical structure. Writhe is a measure of the overall folding of the DNA molecule and is the most readily apparent change in DNA that accompanies supercoiling.

Supercoiling puts DNA under a great deal of stress—at a σ value of -0.06, the DNA is on the brink of local unwinding, in which base pairs are broken. The selective advantage of negative supercoiling is that it greatly favors any process that unwinds DNA, such as replication, transcription, repair, and recombination.

Supercoiling, in fact, does more for DNA than act as an executive enhancer; it keeps the unruly, spreading DNA inside the cramped confines that the cell has provided for it. This space crunch came about because a long time ago, nature took the easy way out and evolved a linear genetic code. It might have been better for the complex contemporary organisms if a two- or three- dimensional code had evolved. But once the choice of a linear code was made, it was irreversible. The result is that DNA is much longer than the space it is forced to occupy. So one of the major functions of the supercoiling of DNA in chromatin is to compact DNA in an orderly, easily accessible fashion. The supercoiling of DNA around the core nucleosomes of chromatin has been studied extensively (Richmond et al., 1984). The energy necessary for the tight supercoiling of DNA in nucleosomes is provided by the binding of the DNA to histones. The highly ordered supercoiled DNA in chromatin is, however, largely inert. The energy of protein-driven supercoiling cannot be used to power the opening of the double helix without losing the energy of the DNA–histone interaction. The regular supercoiled DNA in chromatin is likewise inhibited from attaining the variety of conformations that promote the long-range interaction between different DNA sites.

The active form of supercoiling that is maintained by a negative ΔLk is required by many reactions in vivo. I have described proteins that require the conformational or energetic assistance of supercoiling in order to function, as "supercoiling parasites" (Cozzarelli, 1980). We have been studying the mechanism of two of these supercoiling parasites—the site-specific recombination enzymes resolvase and Gin (Hatfull and Grindley, 1988; Kanaar and Cozzarelli, 1992). For these enzymes, the old notion of DNA as a kind of generic molecular coat rack, upon which a number of specific DNA binding proteins can be hung, is completely wrong. Instead, the structure of the supercoiled DNA substrate is as important a determinant of the specificity of recombination as is the structure of the recombinase itself (Wasserman and Cozzarelli, 1986). We realized that we needed to understand the conformation of supercoiled DNA in order to understand the mechanism of site-specific recombination. To my surprise, however, there was very little known three or four years ago about the conformation of supercoiled DNA in solution, primarily because it is impossible to study supercoiling structure directly. High-resolution techniques, such as X-ray crystallography and nuclear magnetic resonance, are inapplicable, because supercoiled DNA is too big and irregular. Hydrodynamic methods, such as sedimentation, end up telling almost as much about the ingenuity of the experimenter as about supercoiling structure.

We decided to take a gamble on an indirect approach to the structure of supercoiled DNA in solution (Boles et al., 1990). We began by modeling negatively supercoiled DNA as a regular interwound helix that was occasionally branched (Fig. 1). The advantage of this structural simplification is that, excluding the branch points, only two geometric parameters of the superhelix need to be measured in order to calculate all the other properties that we were interested in. The work required a collaboration with a geometrician from UCLA, Jim White, and the experiments were carried out at Berkeley by Chris Boles, a former postdoctoral fellow.

We measured, as a function of σ, the length of the superhelix axis and the number of superhelical turns (the number of times the DNA winds around the superhelix axis) (Fig. 1). We used electron microscopy to measure the length of the superhelix. Because the superhelix is long and thin, the distortion in length upon deposition on the microscope grid should be minimal. To measure the number of superhelical turns, we used a topological method that takes advantage of a property of the site-specific recombination enzyme called Int. The number of superhelical turns in any given molecule is not a topological invariant. However, Int converts this geometric property into a topological property—the number of links in a catenane or the number of crossings of a knot—which can be measured accurately by agarose gel electrophoresis.

In order to calibrate the gels, we had to be able to determine the topology

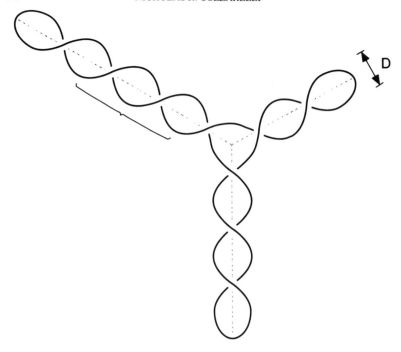

38 NICHOLAS R. COZZARELLI

Fig. 1. Model of plectonemically supercoiled DNA. The DNA is depicted in a plectonemic (interwound), negative superhelical conformation. The superhelix is idealized so that the winding angle is constant. The superhelix diameter, D, and axis (dashed line) are shown. The bracket extends over one superhelical turn. The molecule has one branchpoint.

of catenanes and knots. To do this, we thickened DNA by coating it with *Escherichia coli* RecA protein before viewing by electron microscopy. This allowed us to trace unambiguously the path of the DNA in catenanes or knots. Figure 2 shows examples of micrographs of RecA-coated catenanes.

Fig. 2. Electron microscopy of catenated DNA. The catenanes were isolated from a bacterial mutant that accumulates catenated intermediates in DNA replication. The DNA was denatured, coated with RecA protein, and visualized by electron microscopy. The protein-coated DNA is approximately 100 Å thick and allows the discrimination of the overpassing and underpassing stands at crossings (nodes). Below each micrograph is a drawing of the catenane. **a–c:** 4-noded catenanes; **d–e:** 6-noded catenanes; **f–g:** 8-noded catenanes; **h:** 10-noded catenane; **i:** singly linked catenane in which one of the rings is a knot. The molecule in (h) appears to be about 2.5 times longer than those in the other panels because a different protein-coating method was used.

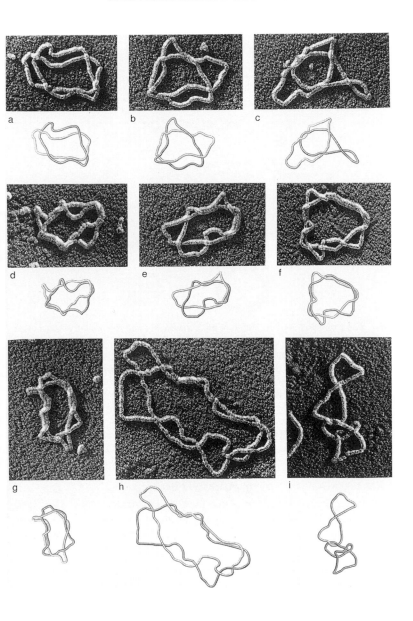

Figure 2.

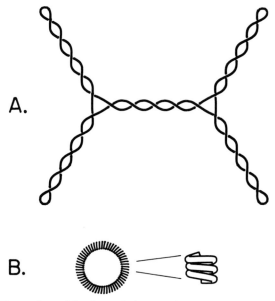

Fig. 3. Comparison of the shape of plectonemically and solenoidally supercoiled DNA. **A:** A diagram of a 4.6 kb plectonemically supercoiled DNA molecule with a σ of −0.06. The DNA is modeled as a regular right-handed superhelix except for the deviations at the ends of the superhelix axis and the branch points. The line representing the DNA double helix in (A) has a width of 22 Å in the scale of the figure. **B:** To compare the geometry of plectonemic supercoils with a model for the supercoiling found in nucleosomes, the same 4.6 kb DNA is wound into a smooth left-handed solenoidal superhelix with a radius of 43 Å and a pitch of 28 Å. (Data derived from Richmond et al., 1984.) The solenoidal model has a σ of −0.079. The scale in (B) is the same as in (A), except that, for clarity, the diameter of the double helix has been reduced by 50%; a blowup of the supercoils is also shown.

We then calculated the writhe and twist of supercoiled DNA, as well as the superhelix diameter and winding angle (Fig. 1). We were led to a model of supercoiled DNA in solution that is shown in Figure 3A. In contrast, Figure 3B shows a DNA of the same length and nearly the same σ that contains nucleosome-like solenoidal supercoils. The difference between the molecules is striking. The nucleosome-like form of supercoiling is a great way to compact DNA, but the free form promotes the dynamic interaction of distant DNA sequences.

There were several limitations, however, to the results presented thus far. First, our conclusions were based on reasonable but unproven assumptions, so we needed independent conformation of the results. Second, we had to use a perfectly regular model for supercoiling to do the calculations, but an important aspect of negative supercoils in solution is the dynamic variation of structure about mean values. We knew nothing of these distributions. Third, we could not explain some of the properties we found on an energetic basis, because none of our measurements were directly related to the thermodynamics of supercoiling.

All these limitations were overcome by a Monte Carlo simulation of DNA supercoiling (Vologodskii et al., 1992). This work was done in collaboration with experts on computer simulations of DNA: Alexander Vologodskii, Konstantin Klenin, and Maxim Frank-Kamenetskii from Moscow. In Berkeley, the calculations were done by Steve Levene, then a postdoctoral fellow, who is also an expert in these techniques. Much of the work was done when Alex spent about eight months with us in Berkeley.

Not many outside the realm of physical chemistry know how Monte Carlo simulations are done, so I will start by outlining the procedure. It is an iterative approach to an equilibrium set of conformations. We used a simplified model for DNA, called "the discrete wormlike chain," in which closed DNA is modeled as a polygon of short cylindrical segments. The length of the cylinders is unimportant, as long as they are short enough for the molecules to approximate a wormlike coil, and we met this criterion. The diameter of the cylinders is the effective diameter of DNA and is usually greater than the geometric diameter of DNA because of electrostatic repulsion between the backbone phosphates. In a separate study (Rybenkov et al., 1993), we computed the effective DNA diameter as a function of NaCl concentration. We imparted to the model DNA chains a certain length and σ value, and the literature values for the energy it takes to twist and bend DNA. With this information, it is easy for a computer to calculate the elastic energy of the chain. The computer then deforms the chain a little, either by rotation of a small segment of the chain about some angle or by a slithering of a portion of the chain. This generates a trial conformation.

If the energy of the trial conformation is less than that of the previous conformation, the move is always accepted. If the energy is greater, the move is accepted with a probability dependent on the energy difference between trial and prior conformations, according to a fundamental algorithm developed by Metropolis and associates in 1953. An example of the early

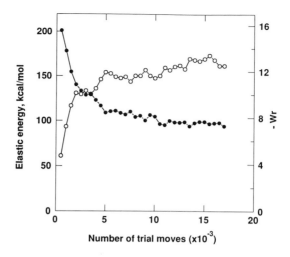

Fig. 4. Dependence of the elastic energy (●) and writhe, *Wr*, (○) of a DNA chain on the number of trial moves in a Monte Carlo simulation. The chain was 3.5 kb long and had a σ of −0.05. Each point represents an average of over 500 successive trial moves.

course of a simulation is shown in Figure 4, in which the negative writhe and the elastic energy of the chain are plotted as a function of the number of trial moves. Both parameters asymptotically approach a limiting range of values, within which they fluctuate statistically. An important feature of the Monte Carlo procedure is that it gives not a single minimum energy conformation but an ensemble of conformations that, in theory, reflects the population of DNA conformations in solution. In practice, we threw away the first million moves of each run and saved the next ten million for data analysis.

Figure 5 shows stereo drawings of conformations we obtained at four different values of σ. The length of the DNA was 3.5 kb and the effective helix diameter was 35 Å, its value at high electrolyte concentrations. When relaxed, the DNA molecule had a rather floppy open structure (Fig. 5A). But, as soon as σ reached −0.03 (Fig. 5B) or less, the DNA displayed an interwound (plectonemic) structure. We never saw solenoidal superhelices. As the supertwist density increased, the molecules became more regular and skinnier (Figs. 5C and 5D).

We established by both quantitative and qualitative means that the simu-

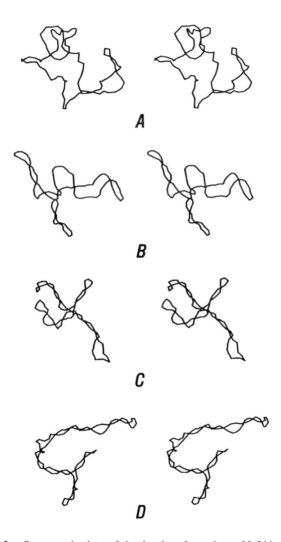

Fig. 5. Stereoscopic views of simulated conformations of 3.5 kb supercoiled DNA molecules. **A:** $\sigma = 0$; **B:** $\sigma = -0.03$; **C:** $\sigma = -0.05$; **D:** $\sigma = -0.07$. The superhelices in (B), (C), and (D) each have a single branchpoint.

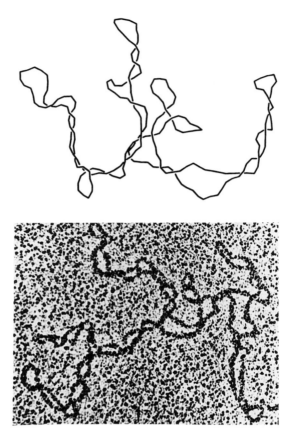

Fig. 6.   Comparison of an electron micrograph of a 7 kb supercoiled DNA molecule with a simulated chain of identical length and superhelix density ($\sigma = -0.045$).
The simulated DNA (above) was flattened to mimic the effect of spreading for microscopy. The electron micrograph is below.

lated molecules were good representations of supercoiled DNA molecules in
solution. In a qualitative test of validity we asked the computer to gently
flatten simulated supercoiled DNA conformations, a procedure which mimics the spreading of a DNA molecule on an electron microscope grid (Fig.
6). Below this simulated flattened conformation in Figure 6 is an electron
micrograph of a supercoiled DNA molecule of the same length and σ value.

The calculated and experimental molecules look strikingly similar. They are long, thin, regular, plectonemic, branched structures.

Quantitatively, the agreement between the experimental data and the simulated data is also excellent, actually better than I expected. In fact, there is no difference between the simulated and experimental data within the errors of the methods.

One quantitative test of agreement was the number of superhelical turns as a function of ΔLk (Fig. 7). The open circles are the experimental data from the Int probing experiments I mentioned previously, and the filled symbols are the simulated data. Both sets of data fit to the same straight line, and indicate that the number of superhelical turns is a linear function of ΔLk.

Figure 8 shows the length of the superhelix axis as a function of σ. The closed symbols are the simulated data, and the open symbols are the electron microscopy measurements. Once again the data are, within error, the same for both methods, and show that the length of the superhelix axis is independent of σ. This means that as supercoiling increases, the superhelix

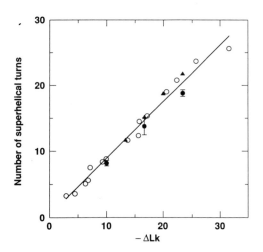

Fig. 7.   Comparison of the simulated and experimental values for the average number of superhelical turns as a function of ΔLk. The data are for 3.5 kb molecules. Two separate procedures were used to determine the Monte Carlo values (●, ▲). The experimental data (○) of Boles and coworkers (1990) were obtained from Int recombination assays. The solid line indicates the linear dependence of the number of superhelical turns on ΔLk. Error bars indicate one standard deviation.

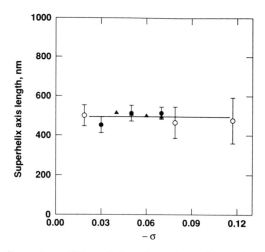

Fig. 8.   Comparison of Monte Carlo and experimental determinations of the length of the superhelix axis as a function of superhelix density. The DNA was 3.5 kb. The Monte Carlo results were obtained using two procedures (●, ▲), and the experimental values (○) are the electron microscopy results of Boles and coworkers (1990). The solid line shows the average value of the superhelix axis length determined from all the data. Error bars indicate one standard deviation.

is unchanged in length but becomes skinnier. This length constancy maintains the winding angle of the superhelix at about 54°, a result which had been predicted on theoretical grounds by Camerini-Otero and Felsenfeld (1978).

The data from topological methods, electron microscopy, and computer simulation all agree. Since these methods have very different assumptions and limitations, their agreement gives us confidence in the validity of the results. Thus, we believe that the following points are reasonably well established: First, supercoils in solution are plectonemic and frequently branched. Second, supercoiling changes not only the writhe of the DNA but also the twist and the helical repeat. The ratio of the changes in writhe and twist is nearly independent of σ, and equals about 3. Third, the superhelix axis length is another constant, and equal to about 40% of the DNA length. This emphasizes how long and thin the plectonemic superhelix is, because the worst possible job of compaction of a circular molecule gives a superhelix axis length 50% of the DNA length. Fourth, the number of supercoils is a linear function of ΔLk, and the proportionality constant is 0.9.

The simulations did more than confirm our earlier results: they told us a great deal about supercoiling in areas in which there was little or no experimental data. For example, using an algorithm that counts the number of branch points for every conformation, we found that branching of the superhelix incurs a heavy enthalpic cost. At a σ value of -0.06, the enthalpy change is about 5 kcal/mol for the introduction of a single branch into a superhelical DNA molecule. This reflects the energy needed to put in the tight hairpin turn at the end of the superhelix branch and to bend the DNA at the branch point. Branching in supercoiled DNA occurs because it is entropically favorable, as it permits many more chain conformations.

Many proteins involved in transcription, replication, and recombination bend the DNA that they bind to (Nash, 1990; Echols, 1990). These proteins will promote branching, because the energy of binding pays a portion of the enthalpic cost of branching (Kanaar and Cozzarelli, 1992). Such proteins are expected to be preferentially located at the apices of the hairpin turns of superhelical DNA or at the branch points. There is emerging support for this expectation in the case of transcription complexes with *E. coli* RNA polymerase (Ten Heggeler-Bordier et al., 1992) and kinetoplast bend sequences (Laundon and Griffith, 1988). The apical location is important in function— for example, it helps to explain the activating effect of the catabolite activator protein on transcription (Zinkel and Crothers, 1991). Moreover, a number of recombination reactions require the interaction of three separate sequences, and these sequences are located preferentially at branch junctions (Johnson, 1991; Kanaar and Cozzarelli, 1992).

There is great variation within and among experimental studies on superhelix branching. The simulation data help to explain why. They show that branching is a sensitive function of many factors—including the length of the DNA, the value of σ, and the salt concentration—which are usually not controlled in separate investigations. I used to think of the branching of a superhelix as analogous to the branching of the rubber band that powers a model airplane, but this turns out to be a misleading comparison. The branching of the rubber band increases directly with torque, but there is an inverse relationship between superhelix density and branching. The simulations explain the latter relationship. As superhelix density increases, the apical hairpin turn tightens, and the enthalpic cost increases.

Analogous arguments explain why branching increases with diminishing salt concentration and increasing DNA length. The effective helix diameter increases as the shielding cation concentration decreases and bends are forced to be more gentle. Increasing DNA length augments the entropic gain of branching.

We found that the relative contributions of enthalpy and entropy to the free energy of supercoiling are not independent of σ. At a low superhelix density, enthalpy directs the whole show. At physiological levels of superhelix density, enthalpy and entropy contribute equally. The floppy chain at a low superhelix density carries with it little entropic cost, but the entropic term increases as supercoiling limits chain conformations.

An important biological aspect of supercoiled DNA is its promotion of the protein-mediated interaction between distant DNA sites. Generally, a bivalent protein or protein complex will interact more readily with two sites in *cis* (on the same DNA molecule) than with two in *trans* (on two different DNA molecules). We calculated the local concentration of two sites in *cis*, that is, the concentration of two sites in *trans* that have the same probability as the sites in *cis* of being within some small fixed distance of each other. For two sites 300 bp apart on a 3.5 kb linear DNA molecule, the local concentration is 80 μg/ml. Since a typical DNA concentration in an enzymology experiment is 5 μg/ml, this is a 16-fold enhancement by the *cis* configuration. If the DNA is cyclized and supercoiled, the local concentration is about 10,000 μg/ml—a super-*cis* effect!

For any given molecule, the extent of the super-*cis* effect does not depend greatly on the contour separation between sites. The superhelix is a dynamic structure, and no matter how far apart two sites are along the chain contour, they will at the same fraction of time be close together in three-dimensional space.

The biological importance of the super-*cis* effect is that two DNA sites are much more likely to interact with a protein if the molecule is supercoiled, and the interaction will be essentially independent of distance. Conversely, if an interaction between two sites is known to fall off with distance, there must be some special mechanism to bring this about. For example, the *cis* effect of a transcriptional enhancer can be limited to suit the physiology of the situation by limiting the conformation freedom of supercoiling.

Computer simulation has been perhaps the most important technique in studying the conformation of DNA supercoils. It has been equally valuable in analyzing the conformation of catenated DNA (Vologodskii and Cozzarelli, 1993). I believe that computer simulation will be increasingly important in biology. As computers become ever faster and cheaper, more complex problems can be addressed. Simulation may become as routine a technique in molecular biology as gel electrophoresis.

I want to switch now to our recent studies of the topology of the kineto-plast DNA network. This work was again a collaboration; the kinetoplast DNA experts were Paul Englund at Johns Hopkins and Carol Rauch, his graduate student. The graph theory depended on Jim White. Most of the experiments I will discuss were done at Berkeley by Junghuei Chen, a post-doctoral fellow. The kinetoplast DNA network may seem to be a very differ-ent topic from supercoiling, but by the end, I hope, you will see that they are closely related.

A series of unicellular eukaryotic parasites are characterized by the pres-ence of a kinetoplast, a highly modified mitochondrion (Ryan et al., 1988). Members of this family of parasites include *Leishmania, Crithidia,* and *Try-panosoma.* The kinetoplast provides the parasite with energy for flagellar mo-tion. The variety of parasites that have a kinetoplast suggests that there is something fundamental about its highly unusual structure. The kinetoplast is distinguished by a catenated network of about five thousand DNA rings that takes up much of the interior space of the kinetoplast. The rings come in two sizes: minicircles and maxicircles. The maxicircles are biologically es-sential, but seem structurally irrelevant. After selective linearization of maxicircles with a restriction enzyme, the network remains intact. The minicircle catenation is the primary determinant of the network structure. We did our work with the *Crithidia* network, which contains 5,000 minicircles about 2.5 kb in size.

When we started our collaboration, I felt confident that with our extensive experience with DNA topology we could penetrate the structure of the ki-netoplast catenanes just by viewing electron micrographs. After a few weeks of work, however, we had gotten absolutely nowhere. We decided that in-stead of trying to solve the seemingly intractable real-life problem, we would simplify the problem to something that we could solve, an approach that had been successful in our study of supercoiling conformations.

Because the kinetoplast network is probably a monolayer of DNA rings (Ryan et al., 1988), we could introduce a tremendous simplification and model the network as a planar graph. (An introduction to graph theory is provided by Grünbaum and Shephard, 1989.) A graph that is planar can be drawn without any edge crossings. A further simplification was to assume that all minicircles have the same structure and connections to other circles. This seems a biologically reasonable assumption, as this elaborate network is pre-sumably made in some regular repeating fashion. We ignored maxicircles, boundaries, holes, or any additional network heterogeneity.

Since we were modeling the network as a regular planar graph, only two numbers needed to be determined to characterize the overall topology of the network. One was the number of links between neighboring rings in the network. We showed by electron microscopy and by gel electrophoresis that this number is uniquely 1 (Rauch et al., 1993). We have not determined the topological sign of the interlocking, but it is probably an equal mixture of + and −.

The other number that characterizes a network is the valence. In the graphs shown in Figure 9, the circles represent the minicircles and the lines represent topological bonds to neighboring rings. The valence is the number of lines emanating from each ring—in physical terms, it is the number of rings that each ring is linked to. Graph theory simplifies this problem enormously, because it shows that the valence of a uniquely tiled planar graph can only be 3, 4, or 6. This result was very important psychologically for us, because we had started with an intractable problem and were left with an easy one.

We determined the valence by a statistical method. A kinetoplast DNA network was randomly attacked with a restriction enzyme, and the degradation products were analyzed by gel electrophoresis. The fragmentation pattern depends upon the valence of the network. The amount of full-length linears, dimeric catenanes, and trimeric catenanes were measured as a function of digestion. The pattern of monomers, dimers, and trimers all showed that the valence of the network is 3.

The minicircles released by partial degradation were all relaxed and covalently closed. We also showed that the interlocking of the rings into the network is done in a strain-free manner that causes no writhing of the minicircles (Rauch et al., 1993).

For the rest of this article, I will speculate freely on what these results might mean. I believe that the findings that the minicircles are relaxed and singly interlocked with a valence of 3 suggests a great deal about network structure and replication. Establishment of the validity of these speculations and suggestions awaits experimental tests.

In Figure 10 is a to-scale model of a portion of the kinetoplast DNA network, rather than the graphical representation in Figure 9. The straight lines delineate the unit cells and emphasize the repeating pattern of the network. Because the actual kinetoplast network is finite, we relaxed our assumption of uniform minicircle connections at the edge of the network. In Figure 10, the valence is 2 for the outermost rings. It is thought that replication of the kinetoplast DNA network proceeds in the following stages (Ryan et al., 1988): A covalently closed minicircle is removed from the network by a topoisomerase and replicated. The gapped products of replication are reattached to the net-

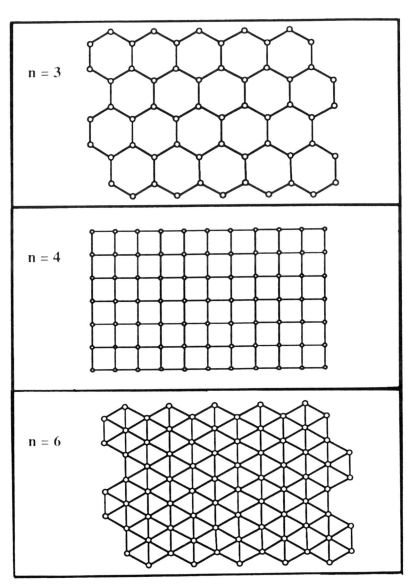

Fig. 9. Examples of regular planar graphs. Shown are diagrams of representative portions of infinite planar graphs, in which the small circles are the vertices and the lines connecting them are the edges. The diagrams model possible structures for the kinetoplast DNA catenated network in which the vertices represent minicircles and the edges the topological bonds between them. The value n is the valence of the graphs. The graphs shown contain hexagonal ($n = 3$), square ($n = 4$), and triangular ($n = 6$) tiles.

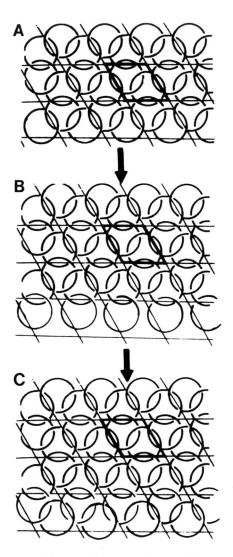

Fig. 10. A speculative model for the replication of the kinetoplast DNA network. Shown is a small section of a network including its boundary at the bottom. The network is ruled off into unit cells. **A:** Network before replication. The valence of the network minicircles is 3, except at the boundary, where it is 2. **B:** A row of minicircles is added to the boundary to complete the valence shell of the boundary of the network shown at the top. **C:** A second row of rings is added by linking to neighboring pairs of the new rings added in B. The net result is the addition of a row of unit cells to the network.

work periphery by a topoisomerase; then the network is split in half and the gaps are filled in. The simplest scenario for the network growth phase is a random stuffing by topoisomerases. The topoisomerases, we assume, will keep on catenating replicated minicircles to the growing network until the valence shells are filled. Because the edge is the only place with unfilfilled valence shells, it is the only place where the network can grow. The result is the addition of a row of DNA rings to the network periphery (Fig. 10B). Because the valence shell is not complete for the new row of minicircles, topoisomerases can link in still another row. This row can be catenated to both neighboring rings in the previously added row (Fig. 10C). Thus, in two steps a row of unit cells has been added and the basic network topology has been perpetuated by a simple mechanism.

The network, even as a monolayer, has practically filled up the kinetoplast. A monolayer is advantageous for packing, orderly replication, and perhaps also for function. The mathematics of the situation helps explain why the network is a monolayer—that is, why it can be described by a planar graph. A perfectly regular graph with a valence of 2 is a line. A completely regular graph of valence 3 must be planar. A regular three-dimensional graph requires a higher valence. Therefore, limiting the valence to 3 helps limit a graph to a planar form, and the network it models to a monolayer.

The minicircles are singly interlocked to each other in a network. We showed a number of years ago that catenanes made by DNA gyrase are also singly interlocked (Krasnow and Cozzarelli, 1983). More recently we demonstrated this for catenanes made by phage T2 topoisomerase (Adams et al., 1992). The catenanes made by topoisomerases are singly interlocked because there is no enthalpic cost to single interlocking. In contrast, there is a substantial enthalpic cost of a double interlock, because it induces writhe into the constituent rings (Wasserman et al., 1988; Vologodskii and Cozzarelli, 1993). It is reasonable, then, that the kinetoplast network is limited to single interlocking. Single interlocking has the additional feature of maximizing the valence of a given-sized ring. If the rings were doubly interlocked, then the valence would have to be less than 3, and no regular network monolayer would be formed.

The lone exception to the generalization with which I started this essay— that the DNA in all organisms is negatively supercoiled—is the kinetoplast network. In this compartment, the DNA is relaxed (Rauch et al., 1993). I think that this exception is critical to forming a network. Relaxation maximizes the linking space for a given-sized DNA ring. If the DNA were supercoiled, the valence of the network would have to be lower; the kinetoplast DNA would either have large holes or be a linear array. There is direct evidence for the importance of a relaxed topology. Catenanes made by DNA gyrase with supercoiled DNA plasmids were oligomers only, but *mirabile dictu,* gigantic catenated networks were formed with relaxed DNA (Kreuzer and Cozzarelli, 1980).

But why have a kinetoplast network at all (Borst, 1991)? One possibility is that the kinetoplast network is a primitive form of chromatin. Kinetoplasts have a bizarre transcription system in which maxicircles encode error-riddled transcripts that are corrected with the aid of RNAs that can be encoded by the minicircles (Sturm and Simpson, 1990). Therefore transcription can involve a cooperation between maxicircles and minicircles. To prevent loss of maxicircles and minicircles, they are kept together as a catenane. A conventional chromosome keeps genes together by covalent rather than topological links. A chromosome must also compact DNA in an orderly fashion. The total length of the minicircle DNA in a *Crithidia* network is about 4,200 μ. The length of these minicircles catenated into a row would be on the order of 700 μ. Upon making a square two-dimensional network, the edge is reduced to 14 μ. There is a further reduction of the extent of the network *in vivo,* to about 1 μ; presumably, this is brought about by condensing proteins other than histones. The net effect is condensation of minicircle DNA by nearly four orders of magnitude, and that is about the same as the compaction achieved in nuclear chromatin. The kinetoplast network has substituted condensation directed by catenation for that directed by supercoiling.

ACKNOWLEDGMENTS

I am grateful to the National Institutes of Health and to the National Science Foundation for their generous support of my research throughout my career. The computations described would not have been possible without the time on Cray computers provided by the San Diego Supercomputer Center and the Berkeley Computation Center. I have listed in the text the graduate students, postdoctoral fellows, and senior collaborators who did the experiments described. However, the projects were shaped by continual conversations with colleagues and particularly with the other members of my laboratory. It is a pleasure to acknowledge my gratitude to them.

REFERENCES

Adams, D.E., Shekhtman, E.M., Zechiedrich, E.L., Schmid, M.B., and Cozzarelli, N.R. (1992). *Cell* **71**, 277–288.
Bauer, W.R. (1978) *Annu. Rev. Biophys. Bioeng.* **7**, 287–313.

Boles, T.C., White, J.H., and Cozzarelli, N.R. (1990). *J. Mol. Biol.* **213,** 931–951.

Borst, P. (1991). *Trends Genet.* **7,** 139–141.

Camerini-Otero, R.D., and Felsenfeld, G. (1978). *Proc. Natl. Acad. Sci. U.S.A.* **75,** 1708–1712.

Cozzarelli, N.R. (1980). *Science* **206,** 1081–1083.

Cozzarelli, N.R., Boles, T.C., and White, J.H. (1990). *In* ''DNA Topology and Its Biological Effects '' (N.R. Cozzarelli, and J.C. Wang, eds.), pp. 139–184, Cold Spring Harbor Laboratory Press, Cold Spring Harbor, NY.

Echols, H. (1990). *J. Biol. Chem.* **265,** 14697–14700.

Grünbaum, B., and Shephard, G.C. (1989). ''Tilings and Patterns: An Introduction.'' W.H. Freeman and Company, New York.

Hatfull, G.F., and Grindley, N.D.F. (1988). *In* ''Genetic Recombination'' (R. Kucherlapati, and G.R. Smith, eds.), pp. 357–396, American Society for Microbiology, Washington, DC.

Johnson, R.C. (1991). *Curr. Opin. Genet. Dev.* **1,** 404–411.

Kanaar, R., and Cozzarelli, N.R. (1992). *Curr. Opin. Struct. Biol.* **2,** 369–379.

Krasnow, M., and Cozzarelli, N.R. (1983). *Cell* **32,** 1313–1324.

Kreuzer, K.N., and Cozzarelli, N.R. (1980). *Cell* **20,** 245–254.

Laundon, C.H., and Griffith, J.C. (1988). *Cell* **52,** 545–549.

Metropolis, N.M., Rosenbluth, A.W., Rosenbluth, M.N., Teller, A.H., and Teller, E. (1953). *J. Chem. Phys.,* **21,** 1087–1092.

Nash, H.A. (1990). *Trends Biochem. Sci.* **15,** 222–227.

Rauch, C.A., Perez-Morga, D., Cozzarelli, N.R., and Englund, P.T. *EMBO J.* **12,** 403–411.

Richmond, T.J., Finch, J.T., Rushton, B., Rhodes, D., and Klug, A. (1984). *Nature* **311,** 532–537.

Ryan, K.A., Shapiro, T.A., Rauch, C.A., and Englund, P.T. (1988). *Annu. Rev. Microbiol.* **42,** 339–358.

Rybenkov, V.V., Cozzarelli, N.R., and Vologodskii, A.V. (1993). *Proc. Natl. Acad. Sci. U.S.A.* (in press).

Sturm, N.R., and Simpson, L. (1990). *Cell* **61,** 879–884.

Ten Heggeler-Bordier, B., Wahli, W., Adrian, M., Stasiak, A., and Dubochet, J. (1992) *EMBO J.* **11,** 667–672.

Vologodskii, A.V., Levene, S.D., Klenin, K.V., Frank-Kamanetskii, M.D., and Cozzarelli, N.R. (1992). *J. Mol. Biol.* **227,** 1224–1243.

Vologodskii, A.V., and Cozzarelli, N.R. (1993). (In preparation.)

Wasserman, S.A., and Cozzarelli, N.R. (1986). *Science* **232,** 951–960.

Wasserman, S.A., White, J.H., and Cozzarelli, N.R. (1988). *Nature* **334,** 448–450.

Zinkel, S.S., and Crotheres, D.M. (1991). *J. Mol. Biol.* **219,** 201–215.

DISCOVERY OF A TRANSCRIPTION FACTOR THAT CATALYZES TERMINAL CELL DIFFERENTIATION

STEVEN LANIER McKNIGHT

Howard Hughes Research Laboratories, Department of Embryology,
Carnegie Institution of Washington, Baltimore, Maryland

I. INTRODUCTION

THE morphological and physiological complexities of multicellular, metazoan organisms rest on the organized elaboration of specialized cell function. Hepatocytes, for example, synthesize an array of enzymes and secretory proteins that allow the liver to carry out its role in detoxification and modulation of blood sugar level. Keratinocytes, on the other hand, synthesize a set of fibrous proteins that intertwine to form the leathery epithelium of the skin. The phenotypic properties of a cell are thus understood to reflect the distinct array of proteins that it synthesizes. The process of cell specialization typically repesents an end state in the ontogeny of a cellular lineage. This process, referred to as terminal cell differentiation, often coincides with the cessation of mitotic proliferation.

Perhaps the single most important biological paradigm that has been established in this century is the role of differential gene expression in terminal cell differentiation. With the exception of the immune system, cells of varying specialized phenotype maintain the same genetic blueprint. Liver cells, that is, contain the same genes as keratinocytes. Cell differentiation does not entail the ordered parceling out of different genes, but instead is mediated by the selective readout of an otherwise invariant genetic blueprint.

The rate-limiting step in selective gene expression is the process of transcriptional initiation. If a protein is not produced in a cell, it likely reflects the fact that its encoding gene is transcriptionally inert. Experiments conducted during the past decade have begun to elucidate the regulatory elements associated with genes that modulate transcription as a function of cell specialization. These regulatory elements, termed enhancers and silencers,

57

The Harvey Lectures, Series 87, pages 57–68

are encoded in the DNA sequence of a gene and act by binding regulatory proteins termed transcription factors. Some such factors, referred to as repressors, act to inhibit (silence) the expression of genes to which they bind. Transcriptional activator proteins do the opposite, causing an induction of expression of the genes to which they attach.

The discriminative action of transcription factors is reflected in their ability to bind selectively to the appropriate array of target genes. Often this selectivity rests on the ability of a transcription factor to bind directly to DNA in a site-selective manner. Proteins of this class are termed sequence-specific DNA binding proteins. In some important cases, however, the specificity of a transcription factor has been found to rely on its ability to "piggyback" itself onto a regulatory DNA sequence via another sequence-specific DNA binding protein (for example, see the review of Thompson and McKnight, 1992).

The role of regulatory proteins in the control of eukaryotic gene expression was largely anticipated from studies of bacterial gene expression. The pioneering work of Jacob, Monod, Gilbert, and Ptashne brilliantly illuminated the notion of regulatory proteins and regulatory DNA sequences (see M. Ptashne, 1986). The conceptual framework outlined from studies of bacterial gene regulation has proven to be fundamentally sound as the field of molecular biology moved into the more complex problem of gene regulation in metazoan organisms. What we have come to realize is that cell specialization is orchestrated by transcription factors that act in much the same way as bacterial repressor and activator proteins. In this lecture I outline experiments conducted over the past several years on a transcription factor that appears to play a fundamental role in terminal cell differentiation. Whereas this protein, termed CCAAT/enhancer binding protein (C/EBP), shares many of the prototypical features of bacterial regulatory proteins, it has also revealed mechanistic properties and biological activities that were largely unanticipated from previous studies of microbial gene regulation.

II. PURIFICATION AND CHARACTERIZATION OF CCAAT/ENHANCER BINDING PROTEIN

In 1984 Barbara Graves, Peter Johnson, and I began preparing nuclear extracts from rat liver tissue as a starting source for purification of transcription factors. Our strategy was to use the DNAaseI footprinting assay developed by Galas and Schmitz (1978) as a means of detecting and tracking proteins capable of binding specifically to regulatory DNA sequences.

The focus of earlier work in my laboratory had centered on transcriptional regulatory elements associated with viral genes, including the herpesvirus thymidine kinase (TK) gene and Moloney murine sarcoma virus (MSV). Such studies had identified small DNA sequence motifs, termed *cis*-regulatory elements, that were somehow responsible for facilitating positive effects on transcription initiation (McKnight and Kingsbury, 1982; Graves et al., 1985). One such sequence, located upstream from the TK gene and within the MSV long terminal repeat (LTR), contained the pentanucleotide consensus 5'-CCAAT-3'. Another *cis*-regulatory element, located within the enhancer of the MSV LTR and termed the enhancer core homology, consisted of the nonanucleotide sequence 5'-GTGGTAAGC-3'. Based on earlier work concerning bacterial gene expression, we imagined that these *cis*- regulatory DNA sequences might serve as binding sites for transcriptional regulatory proteins.

Crude preparations of rat liver nuclei contained activities capable of binding to the CCAAT pentanucleotide and the enhancer core homology. Graves and Johnson, later joined by Bill Landschulz, used conventional chromatographic methods to purify a heat-stable activity that, paradoxically, bound to both the CCAAT pentanucleotide and the enhancer core homology (Graves et al., 1986; Johnson et al., 1987). The polypeptide specifying the activity, termed CCAAT/enhancer binding protein (C/EBP), was purified to homogeneity, allowing derivation of a partial amino acid sequence 17 residues in length. This information facilitated cloning of a cDNA copy of the gene encoding C/EBP (Landschulz et al., 1988a). Expression of C/EBP in bacterial cells provided compelling evidence that the pure protein product was indeed capable of binding to two different *cis*-regulatory DNA sequences. Two important questions were raised at this point. First, how was it possible that this single polypeptide could bind to two different DNA sequences? Second, was C/EBP actually involved in transcriptional regulation? I will return to each of these questions; however, the historical pattern of events led us in a different direction that warranted immediate attention.

III. Conceptualization of the Leucine Zipper Hypothesis

Upon deducing the conceptually translated amino acid sequence of C/EBP, and logging this sequence into a computer data base of known protein sequences, Landschulz discovered pockets of sequence relatedness between C/EBP and the products of several proto-oncogenes (Landschulz et al., 1988a). A region of 11 identities out of 15 residues was shared between C/EBP and the Myc transforming protein. A smaller region consisting of 7

identical residues out of 10 was shared between C/EBP and the Fos transforming protein.

Almost by accident we had identified the location of the DNA binding
domain within C/EBP. During purification from rat liver extracts the 42 kd
protein was highly susceptible to proteolytic cleavage. Indeed, the partial
amino acid sequence that led to cloning of the C/EBP cDNA was derived
from a 14 kd fragment consisting of the 106 carboxyl terminus residues
of the protein (Landschulz et al., 1988a). Since this fragment retained
full DNA binding activity, it allowed rough localization of the DNA binding
domain. The C/EBP-related regions of the Myc and Fos transforming proteins mapped within the DNA binding domain of C/EBP. We therefore speculated that Myc and Fos might turn out to be sequence-specific DNA binding
proteins, and that the pockets of sequence relatedness shared by the three
proteins might reveal the underpinnings of a protein structural motif common to a new class of DNA binding proteins.

By examining the amino acid sequences of C/EBP, Myc, and Fos,
Landschulz and Johnson noted two important facts. First, the sequences were
entirely free of the α-helix–destabilizing residues glycine and proline. Second, each protein contained a region rich in the basic amino acids arginine
and lysine. Anticipating that C/EBP might rely on an α-helical secondary
structure to perform its function as a sequence-specific DNA binding protein, Landschulz plotted its amino acid sequence on a helical wheel. An unusual degree of helical amphipathy was immediately obvious. One face of
the putative α-helix contained an abundance of charged residues that would
be compatible with an aqueous environment. The opposite helical face contained hydrophobic residues including leucine, valine, and isoleucine. Indeed, leucine residues occurred at an invariant heptad interval covering five
consecutive repeats. Landschulz quickly checked the sequences of Myc and
Fos and discovered the same heptad repeat of leucines.

Putting two-and-two-together we reasoned that the hydrophobic faces
of the putative α-helices would represent a dimerization interface that we
called the leucine zipper (Landschulz et al., 1988b). It was reasoned that the
zipper would act to anneal two polypeptides, bringing into close apposition
regions rich in basic amino acids. The latter component, termed the basic
region, was hypothesized to be the polypeptide determinant responsible for
direct DNA contact. These predictions were tested and substantiated in subsequent studies of C/EBP (Landschulz et al., 1989).

Perhaps the most provocative and unconventional aspect of the leucine zipper model held that two different polypeptide chains could coalesce as a heterodimeric complex. This speculation rested on information gained from studies of Fos and yet another transforming protein, termed Jun, which together were understood to comprise a DNA binding activity termed activator protein 1 (AP1). Elegant work from the laboratories of Curran, Vogt, Franza, and Tjian had shown that Fos and Jun form a heteromeric complex. Antibodies to Fos were observed to coprecipitate the Jun protein, and the mixture of Fos and Jun formed a DNA binding complex that bound more avidly to DNA than either protein alone (Franza et al., 1988; Rauscher et al., 1988a; Rauscher et al., 1988b; Turner and Tjian, 1989). We noted that Jun, like Fos, Myc, and C/EBP, contained an α-helix–permissive region decorated with a strict heptad repeat of leucines. Moreover, this putative leucine zipper of Jun was observed to be preceded by a region rich in basic amino acids. Based on these observations it was predicted that Fos and Jun would join via their respective leucine zippers to form the entity responsible for AP1 DNA binding activity (Landschulz et al., 1988b). This prediction was quickly confirmed by a variety of investigators (Kouzarides and Ziff, 1988; Sassone-Corsi et al., 1988).

In the past several years it has become clear that heterodimerization of eukaryotic transcription factors is commonplace. Extensive evidence has emerged documenting mixed dimers in a variety of classes of transcription factors, including those of the leucine zipper, helix-loop-helix, POU:homeodomain, rel:dorsal, and nuclear hormone receptor families (see McKnight and Yamamoto, 1992). It is safe to say that cross-dimerization of transcription factors was not anticipated from studies of bacterial gene regulation. The biological usefulness of heterodimerization is, from my own perspective, unresolved. Does the process broaden the repertoire of regulatory activities of a finite number of genetically encoded transcription factors in a manner analogous to the combinatorial mixing of light- and heavy-chain immunoglobulins? Alternatively, perhaps obligate heterodimerization is used as a means of stringently specifying regulatory decisions such that a genetic program can only be initiated upon negotiation of two independent levels of informational input?

IV. C/EBP IS A MEMBER OF A SUBFAMILY OF LEUCINE ZIPPER PROTEINS

Vertebrate organisms contain genes encoding at least five different C/EBP-related proteins; these are hereafter termed C/EBPα (Graves et al., 1986; Johnson et al., 1987; Landschulz et al., 1988a), C/EBPβ (Akira et al., 1990; Poli et al., 1990; Chang et al., 1990), C/EBPγ (Roman et al., 1990), C/EBPδ (Cao et al., 1991; Williams et al., 1991) and C/EBPε (Williams et al., 1991) according to their chronological order of discovery. These proteins cross-dimerize in all combinations, yet do not cross-dimerize extensively with other leucine zipper proteins such as Fos and Jun. It thus appears that functional subfamilies or ''pods'' of leucine zipper proteins have evolved (Lamb and McKnight, 1991). The restricted patterns of heterodimerization may provide clues as to the interrelated roles and ultimate usefulness of these pods.

V. C/EBP IS CAPABLE OF ACTIVATING TRANSCRIPTION

The assay used to track purification of C/EBPα, DNAaseI footprinting, fell short of establishing a functional role for this protein in transcriptional regulation. Given its propensity to bind avidly to *cis*-regulatory DNA sequences required for positive control of gene expression, we anticipated that C/EBPα would turn out to be a transcriptional activating protein. Evidence supportive of this hypothesis emerged from assays wherein an expression vector capable of producing C/EBPα in mammalian cells was cotransfected into cultured hepatoma cells along with a target gene bearing an avid C/EBPα binding site.

The target gene employed in such studies was the gene encoding serum albumin. Careful work by the laboratories of Schibler and Yaniv had documented the presence of a C/EBPα binding site within the promoter of the albumin gene (Lichsteiner et al., 1987; Cereghini et al., 1987). Importantly, mutations introduced into the albumin promoter that eliminated C/EBPα binding also reduced the transcriptional activity of the promoter as assayed both *in vivo* and *in vitro*. Hepatoma cells were used as a recipient cell for two reasons. First, they were anticipated to contain an appropriate array of regulatory factors necessary for transcription of the albumin gene. Second, relative to adult liver, they were observed to contain abnormally low levels of endogenous C/EBPα. Robust transcriptional activation was observed when the two plasmids, the C/EBPα expression vector and the albumin target, were cotransfected into hepatoma cells (Friedman et al., 1989). The validity of the activating potential of C/EBPα was substantiated by a variety of experimental controls, including the use of a mutated albumin promoter lacking the C/EBPα binding site, as well as mutated variants of C/EBPα that were incapable of binding DNA. Concurrent work by Lane and colleagues demonstrated the transcriptional activating potential of C/EBPα in transfected adipoblast cells (Christy et al., 1989) and in cell-free transcription assays prepared from adipoblasts (Cheneval et al., 1991).

The aforementioned assays provided a means of identifying regions of C/EBPα required for transcriptional activation. Alan Friedman constructed a systematic series of deletion mutants covering a 300-residue region located upstream of the C/EBPα DNA binding domain. By carefully testing the capacity of each mutant to activate albumin transcription in hepatoma cells, two "activating" domains were identified (Friedman et al., 1990). One such domain was localized to a 29-residue segment close to the amino terminus of C/EBPα. The other activating domain was identified covering a larger, centrally located segment of roughly 100 residues. The mechanisms by which the two activating domains function is obscure.

VI. C/EBPα CAN CATALYZE TERMINAL CELL DIFFERENTIATION

It was initially anticipated that C/EBPα would function in an unrestricted manner, perhaps in all mammalian cell types. This expectation was based upon the fact that the *cis*-regulatory motifs to which C/EBPα binds, CCAAT pentanucleotides and enhancer core homologies, are associated with a wide variety of genes transcribed by RNA polymerase II. It has since been realized that many different transcription factors bind to the same *cis*-regulatory motifs as C/EBPα (for review see Johnson and McKnight, 1989). As molecular probes specific to C/EBPα, including cDNAs and specific antisera, became available, it was quickly recognized that the protein is expressed in a substantially restricted manner (Birkenmeier et al., 1989).

Histological staining of adult mouse liver with C/EBP-specific antiserum revealed selective expression in hepatocytes. Other cell types in the liver, including endothelial cells that line ductal and circulatory paths within the liver, appeared to lack C/EBPα. Similarly restricted patterns of expression were observed in other C/EBPα-positive tissues, including skin, intestine, lung, and fat (Z. Cao and S. McKnight, unpublished observations). The pattern of C/EBPα expression in each of these tissues was observed to correlate with terminal cell differentiation. C/EBPα mRNA was observed, for example, in the differentiated keratinocytes of mouse forelimb, but not in the basal cells that replenish the epithelium on a continual basis.

What roles might be anticipated for a transcriptional regulatory protein that functions in a variety of terminally differentiated cell types? One obvious function would entail the activation of cell-type–specific genes such as the albumin gene in hepatocytes or keratin genes in skin cells. Were C/EBPα to participate in cell-type–specific gene activation, it would necessarily require additional informational cues. The presence of C/EBPα in a skin cell does not, for example, cause activation of albumin gene expression. Likewise, C/EBPα does not activate keratin gene expression in hepatocytes. A second hypothetical function for C/EBPα during terminal cell differentiation might involve arrest of mitotic growth. The conversion of stem cells in the basal layer of the skin into mature keratinocytes is accompanied by the abrupt arrest of mitotic growth. The hepatocytes and enterocytes that express C/EBPα are also quiescent with respect to cell division.

We have investigated, in an experimental sense, the hypothetical roles of C/EBPα in cell specialization and growth arrest by the use of cultured 3T3-L1 adipoblasts. Knowing that C/EBPα mRNA was present in adipose tissue, Ed Birkenmeier of the Jackson Laboratory reasoned that 3T3-L1 cells might offer a good system for studying C/EBPα. More than a decade earlier Howard Green had developed the 3T3-L1 cell line as a model system for adipogenesis (Green and Kehinde, 1975). When cultured under appropriate conditions, 3T3-L1 cells proliferate continuously in a blastlike state. Remarkable changes in growth rate, morphology and physiology occur, however, when confluent cultures of 3T3-L1 cells are exposed to an empirically discovered cocktail of adipogenic hormones. Following hormonal stimulation, otherwise fibroblast-like cells proliferate for a brief period, take on a rounded morphology, cease cell division, and assume the phenotype of terminally differentiated adipocytes. The gene encoding C/EBPα is transcriptionally activated during this adipogenic conversion process (Birkenmeier et al., 1989), raising the possibility that its product (C/EBPα) plays a regulatory role in the process of terminal cell differentiation.

The adipogenic conversion process involves the activation of a battery of "fat-specific" genes, including those encoding a variety of fatty acid binding proteins and enzymes important for the synthesis and storage of triglycerides. M. Daniel Lane and his colleagues at Johns Hopkins Medical School have identified and cloned a number of these fat-specific genes. Using transient cotransfection assays, Lane has found that C/EBPα is capable of *trans*-activating the expression of three fat-specific genes (Christy et al., 1989; Kaestner et al., 1990). Each C/EBPα-responsive gene contains a high-affinity C/EBPα binding site within its promoter, and the reasonable assumption is that C/EBPα acts as a direct inducer of the fat-specific genes by binding directly to their promoters, in a manner analogous to its role in activation of albumin gene expression.

Several experiments have provided compelling evidence that C/EBPα does play a fundamental role in the differentiation of adipocytes. Two studies using the ''antisense'' inhibition approach have provided evidence that 3T3-L1 cells cannot differentiate if C/EBPα expression is blocked (Samuelsson et al., 1991; Lin and Lane, 1992). Two complementary studies have provided evidence that premature expression of C/EBPα in otherwise blastlike 3T3-L1 cells catalyzes the differentiation process (Umek et al., 1991; Freytag and Geddes, 1992). The observation that C/EBPα is capable, on its own, of converting 3T3-L1 cells into fully differentiated adipocytes qualifies this protein as a central regulator of terminal cell differentiation. The fact that C/EBPα converts 3T3-L1 cells into adipocytes rather than, for example, into hepatocytes (where it is also expressed) forces the conclusion that undifferentiated 3T3-L1 cells have already negotiated ''determinative'' steps that provide a qualitative blueprint necessary for specifying the terminal action of C/EBPα. This interpretation holds that C/EBPα simply executes a predetermined program of cell specialization, activating an appropriate battery of genes that are somehow imprinted at earlier steps during the specification of a cell lineage.

VII. C/EBPα CAN CAUSE ARREST OF MITOTIC GROWTH

After retrieving a molecular clone of the gene encoding C/EBPα we attempted to express the protein in several different cultured cell lines. An expression vector was constructed wherein the MSV LTR was positioned upstream from the C/EBPα open reading frame (ORF), which was in turn followed by the transcription termination/polyadenylation element of the herpesvirus TK gene. Attempts were undertaken to stably transfect this C/EBPα expression vector into three different cell types; 3T3-L1 cells, human hepatoma cells (HepG2), and LMTK-deficient mouse fibroblasts. In no case was it possible to derive a stably transformed cell line expressing C/EBPα.

Based on these observations, it was provisionally speculated that C/EBPα might regulate mitotic cell growth. In order to further examine this possibility, a conditional form of C/EBPα was prepared by fusing the C/EBPα ORF to the hormone binding domain of the estrogen receptor (ER). Earlier experiments by Yamamoto and colleagues had demonstrated the remarkable ability of the ER hormone binding domain to confer hormone-dependency to otherwise unrelated polypeptides. The E1A protein of adenovirus, for example, was shown to fall under the direct regulatory influence of β-estradiol when fused to the carboxy-terminal segment (350 amino acids) of the human estrogen receptor protein (Picard et al., 1988). Subsequent studies using the Myc protein showed that Myc-ER fusions were also regulated by estrogen (Eilers et al., 1989). Following Yamamoto's lead we prepared a C/EBPα-ER fusion protein, found that its gene regulatory activity was estrogen dependent, and prepared stably transformed 3T3-L1 cells that expressed the C/EBPα-ER fusion protein. Care was taken to remove endogenous estrogenic hormones from the culture medium which was used to maintain the transfected 3T3-L1 cells, including the elimination of phenol red indicator dye which is known to have estrogenic activity.

Stably transfected cell lines expressing the C/EBPα-ER fusion protein cease mitotic proliferation upon exposure to nanomolar concentrations of β-estradiol (Umek et al., 1991). Evidence supporting this finding came from assays of tritiated thymidine incorporation and measurements of cell doubling interval. Whereas activation of the otherwise latent C/EBPα-ER fusion protein arrested cell growth, it did not cause death. Growth-arrested cells expressing the activated fusion protein resume mitotic proliferation upon estrogen removal.

We have yet to determine the molecular basis of C/EBPα's role in growth arrest. Several clues that may be important in interpreting this property of the protein include the observation that growth arrest by C/EBPα is strictly dependent on all three molecular activities of the protein, including its capacity to dimerize via the leucine zipper, bind DNA via the basic region, and activate transcription via its two activating domains (Umek et al., 1991). It would appear, therefore, that the role of C/EBPα in growth arrest might require its ability to activate gene expression. A second observation relevant to the hypothetical role of C/EBPα in growth arrest has been made independently by my former postdoctoral associate, Alan Friedman. In studies of a pluripotent myeloid cell line termed 32D, Friedman has found copious amounts of C/EBPα protein (A. Friedman, personal communication). Assuming that the C/EBPα produced in 32D cells is functionally intact, it would appear that the ability of C/EBPα to impede mitotic proliferation is cell-type–specific. Perhaps C/EBPα is responsible for activating the transcription of genes encoding negative growth factors in certain cell types but not in others. If so, it will be exciting to pursue the nature of such factors and to determine why their response to C/EBPα is cell-type–specific.

VIII. Perspectives

By focusing our efforts on an obscure transcriptional regulatory protein, my colleagues and I have stumbled across several interesting observations that may have broad biological relevance. Studies of C/EBPα and its relationship to the Fos, Myc, and Jun transforming proteins led to the idea that that transcription factors can be composed of distinct polypeptide chains. The underlying resource of this ordered and sometimes obligate mixing of distinct polypeptide chains is unclear. It has become clear, however, that heterotypic mixing of transcription factors in metazoan organism is a widespread phenomenon—perhaps constituting the rule rather than an exception to the repressor paradigm established from studies of bacterial gene regulation.

Studies of C/EBPα have also provided unanticipated insight into the process of terminal cell differentiation. Starting with the observation that expression of C/EBPα is concordant with terminal differentiation in several different tissue types, it has been possible to demonstrate both the necessity and sufficiency of this single protein to catalyze the final step of adipocyte differentiation. It is clear that, in a qualitative sense, C/EBPα cannot act alone to specify the differentiated phenotype of a cell. Its role instead appears to be dedicated to the terminal execution of otherwise predetermined genetic programs characteristic of various tissues including liver, lung, intestine, skin, and fat.

Finally, the cell types of many tissues in which C/EBPα is expressed are growth arrested. We have discovered that C/EBPα directly specifies growth arrest in cultured 3T3-L1 fibroblasts. Does C/EBPα undertake such a role *in vivo*? Moreover, if so, by what mechanism does it block mitotic proliferation? These latter questions, by challenging the validity of past studies, should provide valuable focal points for future studies on C/EBPα.

REFERENCES

Akira, S., Isshiki, H., Sugita, T., et al. (1990). *EMBO J.* **9,** 1897–1906.

Birkenmeier, E.H., Gwynn, B., Howard, S., Jerry, J., Gordon, J.I., Landschulz, W.H., and McKnight, S.L. (1989). *Genes Dev.* **3,** 1146–1156.

Cao, Z., Umek, R.M., and McKnight, S.L. (1991). *Genes Dev.* **5,** 1538–1552.

Cereghini, S., Raymondjean, M., Carranca, A.G., Herbomel, P., and Yaniv, M. (1987). *Cell* **50,** 627–638.

Chang, C.-J., Chen, T.-T., Lei, H.-Y., Chen D.-S., and Lee, S.-C. (1990). *Mol. Cell. Biol.* **10,** 6642–6653.

Cheneval, D., Christy, R.J., Geiman, D., Cornelius, P., and Lane, M.D. (1991). *Proc. Natl. Acad. Sci. U.S.A.* **88,** 8465–8469.

Christy, R.J., Yang, V.W., Ntambi, J.M., Geiman, D.E., Landschulz, W.H., Friedman, A.D., Nakabeppu, Y., Kelly, T.J., and Lane, M.D. (1989). *Genes Dev.* **3,** 1323–1335.

Eilers, M., Picard, D., Yamamoto, K.R., and Bishop, J.M. (1989). *Nature* **340,** 66–68.

Franza, B.R., Jr., Rauscher, F.J., III, Josephs, S.F., and Curran, T. (1988). *Science* **239,** 1150–1153.

Freytag, S.O., and Geddes, T.J. (1992). *Science* **256**, 379–382.

Friedman, A.D., Landschulz, W.H., and McKnight, S.L. (1989). *Genes Dev.* **3**, 1314–1322.

Friedman, A.D., and McKnight, S.L. (1990). *Genes Dev.* **4**, 1416–1426.

Galas, D., and Schmitz, A. (1978). *Nucleic Acids Res.* **5**, 3157–3170.

Graves, B.J., Eisenman, R.N., and McKnight, S.L. (1985). *Mol. Cell. Biol.* **5**, 1948–1958.

Graves, B., Johnson, P.F., and McKnight, S.L. (1986). *Cell* **44**, 565–576.

Green, H., and Kehinde, O. (1975). *Cell* **5**, 19–27.

Johnson, P.F., Landschulz, W.H., Graves, B.J., and McKnight, S.L. (1987). *Genes Dev.* **1**, 133–146.

Johnson, P.F., and McKnight, S.L. (1989). *Annu. Rev. Biochem.* **58**, 799–839.

Kaestner, K.H., Christy, R.J., and Lane, M.D. (1990). *Proc. Natl. Acad. Sci. U.S.A.* **87**, 251–255.

Kouzarides, T., and Ziff, E. (1988). *Nature* **336**, 646–651.

Lamb, P., and McKnight, S.L. (1991). *Trends Biochem. Sci.* **16**, 417–422.

Landschulz, W.H., Johnson, P.F., Adashi, E.Y., Graves, B.J., and McKnight, S.L. (1988a). *Genes Dev.* **2**, 786–800.

Landschulz, W.H., Johnson, P.F., and McKnight, S.L. (1988b). *Science* **240**, 1759–1764.

Landschulz, W.H., Johnson, P.F., and McKnight, S.L. (1989). *Science* **243**, 1681–1688.

Lichtsteiner, S., Wuarin, J., and Schibler, U. (1987). *Cell* **51**, 963–973.

Lin, F.-T., and Lane, M.D. (1993). *Genes Dev.* (in press).

McKnight, S.L., and Kingsbury, R. (1982). *Science* **217**, 316–324.

McKnight, S.L., and Yamamoto, K.R. (1993). ''Transcriptional Regulation.'' Cold Spring Harbor Press, Cold Spring Harbor, NY, (in press).

Picard, D., Salser, S.J., and Yamamoto, K.R. (1988). *Cell* **54**, 1073–1080.

Poli, V., Mancini, F.P., and Cortese, R. (1990). *Cell* **63**, 643–653.

Ptashne, M. (1986). ''A Genetics Switch.'' Blackwell Scientific Publications, Palo Alto, CA.

Rauscher, F.J., III, Voulalas, P.J., Franza, B.R., Jr., and Curran, T. (1988a). *Genes Dev.* **2**, 1687–1699.

Rauscher, F.J., III, Cohen, D.R, Curran, T., Bos, T.J., Vogt, P.K., Bohmann, D., Tjian, R., and Franza, B.R., Jr. (1988b). *Science* **240**, 1010–1016.

Roman, C., Platero, J.S., Shuman, J.D., and Calame, K. (1990). *Genes Dev.* **4**, 1404–1415.

Samuelsson, L., Stromberg, K., Vikman, K., Bjursell, G., and Enerback, S. (1991). *EMBO J.* **10**, 3787–3793.

Sassone-Corsi, P., Lamph, W.W., Kamps, M., and Verman, I.M. (1988). *Cell* **54**, 553–560.

Thompson, C., and McKnight, S.L. (1993). *Trends Genet.* (in press).

Turner, R., and Tjian, R. (1989). *Science* **243**, 1689–1694.

Umek, R.M., Friedman, A.D., and McKnight, S.L. (1991). *Science* **251**, 288–292.

Williams, S.C., Cantwell, C.A., and Johnson, P.F. (1991). *Genes Dev.* **5**, 1553–1567.

PARENTAL IMPRINTING IN THE MOUSE

SHIRLEY M. TILGHMAN

Howard Hughes Medical Institute and Department of Molecular Biology,
Princeton University, Princeton, New Jersey

I. INTRODUCTION

A LONG-STANDING assumption of Mendelian genetics was that autosomal genes, present in two copies in diploid organisms, were functionally equivalent. That this is not always the case in mammals has now been firmly established with the discovery of the phenomenon of parental or genomic imprinting. It is now clear that at least a small subset of mammalian autosomal genes are inherited in differentially active states. The decision to be active or silent is determined by the parent from which the gene is inherited. That is, some genes are expressed exclusively from the paternal chromosome and others from the maternal chromosome. One important consequence of parental imprinting is to render the organism functionally hemizygous for the imprinted gene.

The functional nonequivalence of the maternal and paternal genomes in mammals was first suspected from the failure to observe parthenogenesis in mammals (Graham, 1974; Kaufman, 1983). Parthenogenotes are derived from unfertilized eggs whose haploid nuclei have been duplicated by physical or chemical means. Thus their genomes are entirely maternally derived. However several scenarios, in addition to the absence of paternally active genes, could explain the failure of parthenogenotes to thrive, for example the presence of homozygous lethal genes or the absence of some nongenomic contribution from sperm.

The elegant nuclear transplantation studies pioneered by McGrath and Solter (McGrath and Solter, 1983; McGrath and Solter, 1984) were able to distinguish among these possibilities and confirm that both maternal and paternal genomes are required for development. In these experiments, one of the two pronuclei was removed from a fertilized egg, and replaced with a pronucleus

69

The Harvey Lectures, Series 87, pages 69–84

from a second. When the reconstituted embryo contained pronuclei derived from both a male and a female, normal development ensured. However when both pronuclei were derived from either males or females, the resulting embryo failed to thrive. In the case of the entirely female-derived embryo, or gynogenote, the failure was largely in extra-embryonic tissues. With androgenotes, reconstituted from two male pronuclei, it was the embryo proper which was underdeveloped.

The apparent reciprocal nature of the phenotypes exhibited by androgenetic and gynogenetic embryos superficially suggest that the maternal genome plays a more central role in extra-embryonic tissue development, while the paternal genome is more important for the development of the embryo proper. However, several considerations must temper such a simple interpretation. First, the phenotypes observed represent the time in development when at least one imprinted gene product is first required at the correct dosage. Thus one is only observing the first consequence of a disruption in gene dosage, and later consequences cannot be discerned. Second, it has been assumed by many that the defects exhibited in uniparental embryos are the consequence of loss of function of one or more gene products. However, it is just as likely that the overproduction of a gene product(s) leads to the observed perturbations in development. While these experiments, and others that followed in their wake (McGrath and Solter, 1986; Surani et al., 1984, Surani et al., 1986), established the requirement for both parental pronuclei and led directly to the notion that genes are inherited in differentially active states, they could not make any predictions about the number of such genes.

II. Genetic Mapping of Imprinted Domains

Classical genetics has played an important role in elucidating the regions of the mouse genome which are likely to harbor imprinted genes. B. Cattanach, A.G. Searle, and C.V. Beechey have pioneered the use of intercrosses between mice heterozygous for balanced translocations, or mice bearing complete Robertsonian translocations in which two chromosomes are fused at their centromeres (Searle and Beechey, 1978; Cattanach and Kirk, 1985; Searle and Beechey, 1985; Cattanach, 1986; Searle and Beechey, 1990). Snell (Snell, 1946) initially proposed that animals which harbor such translocations undergo a significant degree of nondisjunction during meiosis, leading to the generation of unbalanced gametes, which contain uniparental duplications of the chromosomal region on one side of the translocation breakpoint and

corresponding deletions of the other chromosome involved in the translocation (Fig. 1). When these unbalanced gametes pair during fertilization so as
to restore the appropriate autosomal gene dosage, complete functional complementation is often observed. This can be documented by following the

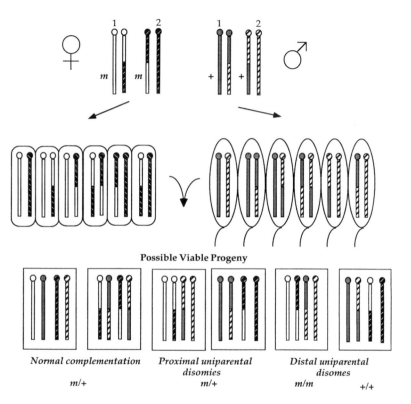

Possible Viable Progeny

Normal complementation

m/+

*Proximal uniparental
disomies*

m/+

*Distal uniparental
disomes*

m/m   *+/+*

Fig. 1. Genetic analysis of genomic imprinting in mice. Males and females carrying a balanced translocation between hypothetical chromosomes 1 and 2 are inter-
crossed. The chromosomes are coded so as to follow separately maternal and paternal
inheritance of each chromosome segment. In the female, chromosome 1 is marked
with a recessive mutation (*m*), while the male is wild type (+). The six possible
gametes that result from nondisjunction during meiosis are drawn on the middle line,
and the potential viable offspring that result from the balanced pairing of these gametes
is drawn on the bottom line. The offspring are either balanced, or carry proximal uniparental disomies or distal uniparental disomies of chromosomes 1 and 2. The predicted phenotype of each class at the *m* locus is indicated at the bottom of the figure.

inheritance of a recessive marker present in only one parent (Fig. 1). In the hypothetical example in Figure 1, the region distal to the translocation breakpoint between chromosomes 1 and 2 can be followed with the recessive marker m, present in the female. If m/m progeny are obtained, then one can conclude that the distal regions of chromosomes 1 and 2 do not contain any imprinted genes. When such progeny are not observed, it is concluded that at least one of the two chromosomes involved in the translocation contains genes which are imprinted.

Using this approach, an ''imprinting map'' of the mouse genome has been generated (Beechey et al., 1990). Over one quarter of the 19 mouse autosomes (chromosomes 3, 4, 13, 15, and 16) can be inherited entirely from one parent without deleterious effect. If imprinted genes exist on these autosomes, gain or loss of their function must have no apparent consequence to the mouse. More likely, they do not contain such genes.

In contrast, strong evidence for noncomplementation has been obtained for regions of chromosomes 2, 6, 7, 11, and 17 (Fig. 2). The proximal segment of mouse chromosome 2, for example, harbors at least one gene which cannot be inherited solely from mothers, as embryos with maternal duplications die early in development. Uniparental inheritance from fathers, however, has no deleterious effect (Cattanach and Kirk, 1985). Mice which inherit the distal third of chromosome 2 exclusively from either mothers or fathers display reciprocal stature defects (Cattanach and Kirk, 1985; Cattanach, 1986; Cattanach et al., 1991). The mice carrying maternal disomies of the region are born with long, flat-sided bodies, while the paternal disomies have short, square bodies. In addition the former mice are apathetic, while the latter display hyperkinetic activity. These phenotypes are consistent with a gain of function of one or more gene products in one instance, and the reciprocal loss in the other, although the apparent reciprocity of the phenotypes could be misleading.

The genetic analysis of chromosome 11 provides a second example of a reciprocal phenotype for maternal and paternal disomies. In this instance, the inheritance of the proximal third of chromosome 11 from mothers leads to smaller-than-normal mice, while inheritance from fathers leads to over sized mice (Cattanach and Kirk, 1985). Once again, the phenotypes can be rationalized by the presence in the region of an imprinted gene, for example a gene regulating growth, whose expression is reduced when inherited from mothers and increased when inherited from fathers.

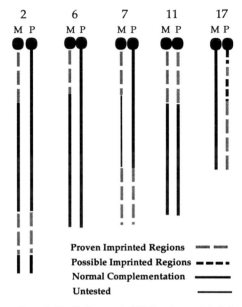

Compiled by B. Cattanach, C.V. Beechey and A.G. Searle

Fig. 2.   Imprinting regions of the mouse genome. Superimposed on the maps of chromosomes 2, 6, 7, 11, and 17 are the regions for which imprinting has been proven, implicated, untested, or discounted. (Reproduced from Beechey et al., 1990, with permission of the publisher.)

III. GENETIC ANALYSIS OF IMPRINTING ON CHROMOSOME 7

Of all the mouse autosomes, chromosome 7 has been the most intensively studied with respect to imprinting. It harbors at least two imprinted domains, and two of the three known endogenous imprinted genes map to the chromosome (DeChiara et al., 1991; Bartolomei et al., 1991). Searle and Beechey (1990) were the first to suggest that chromosome 7 contained imprinted genes. They took advantage of a mouse stock carrying a balanced translocation between 7 and 15 (T(7;15)9H), and the phenotypic marker *albino* (*c*), the gene encoding the enzyme tyrosinase (Fig. 3). *Albino* homozygous mutants can be readily identified by their white coats and lack of pigmentation in the eye. No *albino* progeny were recovered at birth when either fathers or mothers carried the *albino* marker. Those carrying two maternal copies of the distal region of chromosome 7 were retarded in their growth and died between 14

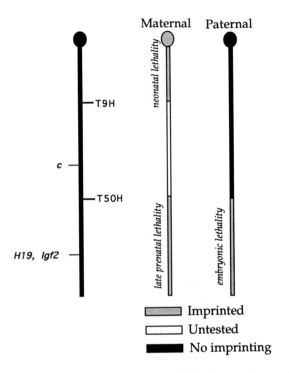

Fig. 3. Imprinting on chromosome 7. On the left, the positions of the genes referred to in the text are displayed on the genetic map of mouse chromosome 7. T9H
and T50H indicate the positions of the translocation breakpoints used to define the
imprinting domains. The mottled regions on the maps of chromosome 7 on the right
identify the chromosomal regions which display either maternal and paternal
noncomplementation. The white area has not been tested as yet, and the black area
displays normal complementation. (Adapted from Searle and Beechey, 1990, with
permission of the publisher.)

and 16 days of gestation. When the fathers carried the *albino* marker, *c/c*
offspring were not detected either at birth or *in utero,* suggesting that the
effect of a paternal disomy of the region was embryonic lethality. From these
and other studies (Searle and Beechy, 1985), the authors concluded that the
distal half of chromosome 7 is a likely site for imprinted genes.

The proximal region of chromosome 7 was also implicated in imprinting,
by virtue of the fact that progeny maternally disomic and paternally deficient
for the region were small at birth and did not survive to weaning (Searle and

Beechey, 1985, 1990). However, the inheritance of two paternal copies of proximal chromosome 7 is not deleterious. Thus both the proximal and distal ends of chromosome 7 display genetic inheritance consistent with the presence of imprinted loci. From the genetics alone one cannot deduce the number of imprinted genes involved in the two domains, nor the direction of the imprints, as the phenotypes could result from either gain of function of one or more gene product(s), a loss of function, or a combination of both.

IV. PARENTAL IMPRINTING OF *INSULIN-LIKE GROWTH FACTOR 2* AND *H19*

The small size of the fetuses carrying maternal duplications of distal chromosome 7 suggested that disruptions in the dosage of imprinted genes led to defects in the control of embryonic and fetal growth. This was borne out by the discovery of the first imprinted gene on chromosome 7, insulin-like growth factor 2 (*Igf2*), a mitogenic polypeptide that is thought to have an important role during embryogenesis. DeChiara, Efstratiadis, and Robertson (DeChiara et al., 1990; DeChiara et al., 1991) generated a mutant mouse strain which had a targeted disruption in the *Igf2* gene. The homozygous mutant progeny were smaller than their wild-type litter mates (approximately 60% of normal body weight). However, *heterozygous* mutant mice also had body weight 60% of normal if the mutation was inherited from fathers, but were normal-sized when the mutation was inherited from mothers. Molecular studies confirmed that the maternal copy of the *Igf2* gene is normally silent, thereby explaining the unusual inheritance pattern of the phenotype. Ferguson-Smith and associates (1991) provided additional evidence in support of the exclusive paternal expression of *Igf2* by showing that embryos carrying a maternal duplication for distal chromosome 7 were deficient in *Igf2* transcripts. This would well explain their small size.

With the discovery of the first imprinted gene, it was now possible to begin a molecular dissection of the phenomenon. The first question one might ask is whether imprinted genes cluster, or whether they are scattered throughout the mouse genome as single genes. The failure to reconcile the phenotype of the homozygous mutant *Igf2* mice (60% of normal size) and the maternal duplication embryos generated in the balanced translocation crosses (late prenatal lethality) led to the conclusion that another imprinted gene must be lurking in the area. This proved to be the case when it was established that the another chromosome 7-specific gene, *H19*, is imprinted as well (Bartolomei et al., 1991).

. The *H19* gene is an unusual gene of no known function. It is transcribed by RNA polymerase II and processed by splicing and polyadenylation, yet it does not appear to encode a protein. No common open reading frame can be discerned in the five mammalian homologs sequenced to date. Yet its nucleotide sequence, and especially its secondary structure, is well conserved in mammals, arguing that it is functional and being selected for in evolution (Brannan et al., 1990; Han and Liau, 1992; Tilghman et al., 1992). Consistent with its presumed noncoding function, *H19* RNA is not associated with polyribosomes, but is located in a conserved 28S particle (Brannan et al., 1990).

Two observations led to the idea that the *H19* gene might be imprinted. First, it had been mapped to the distal third of chromosome 7, close to *Igf2* (T. Glaser and D. Housman, personal communications). Second, when extra copies of the intact gene were introduced into mouse zygotes, the resulting transgenic fetuses died between embryonic day 14 and birth (Brunkow and Tilghman, 1991), that is, at approximately the time when offspring with maternal disomies for distal chromosome 7 were dying.

An RNAase protection assay (Fig. 4a) that could distinguish single base changes between the *H19* gene in different species of *Mus* was exploited to examine the expression of the gene in reciprocal F1 crosses between *M. domesticus,* the species from which most of the standard laboratory strains have been derived, and *M. spretus,* a very distantly related "wild" mouse. Irrespective of the direction of the cross, only the maternally derived allele was expressed (Fig. 4b, lanes 5–11).

Thus *Igf2* and *H19* are imprinted in opposite directions; the *Igf2* gene is expressed from the paternal chromosome and the *H19* gene is expressed from the maternal chromosome. The maternal-specific expression of *H19* can now explain the late prenatal lethality of embryos which contain two maternal copies of the gene, since introduction of extra copies of the gene by transgenic means was lethal at approximately the same time (Brunkow and Tilghman, 1991). It remains to be seen whether the loss of *H19,* together with the gain of *Igf2* is sufficient to explain the earlier lethal effects of inheriting the distal end of chromosome 7 entirely from fathers.

V. GENETIC AND PHYSICAL LINKAGE OF *IGF2* AND *H19*

The *Igf2* and *H19* genes were initially mapped to chromosome 7 using an analysis of somatic cell hybrids which were segregating mouse chromosomes (Pachnis et al., 1984; Lalley and Chirgwin, 1984). An interspecific backcross between *M. domesticus* and *M. spretus* was exploited to examine the

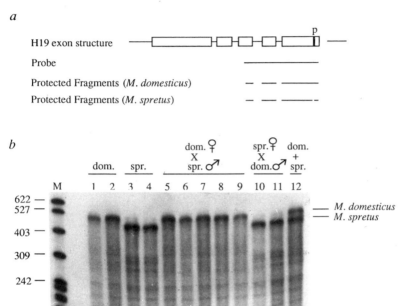

Fig. 4. RNAase protection assay to detect imprinting of the *H19* gene. **a:** The top line of the diagram shows the exon structure of the mouse *H19* gene, drawn in a 5' to 3' orientation. Open boxes indicate exons, and the lines correspond to introns and nontranscribed regions. The filled box in exon 5 and the corresponding ''p'' above the box mark the approximate site of the RNA polymorphism that exists between *M. domesticus* (dom.) and *M. spretus* (spr.). The second line shows a *Bam*HI-*Stu*I 754 bp genomic DNA fragment that was used as the RNAase protection probe. The probe fragments that are protected from RNAase digestion following hybridization by *M. domesticus* or *M. spretus* H19 RNA are drawn in the third line and fourth lines, respectively. **b:** An autoradiogram of RNAase protection assay products of total liver RNA samples from neonatal mice is shown. The F1 progeny of *M. domesticus* females mated to *M. spretus* males (lanes 5–9) and of *M. spretus* females mated to *M. domesticus* males (lanes 10 and 11), as well as the parental lines *M. domesticus* (lanes 1 and 2) and *M. spretus* (lanes 3 and 4) are as indicated. Only the allele-specific protected fragment from exon 5 is included. Marker lane (M) contains radioactively labeled DNA fragments of MspI-digested pBR322, with relative sizes indicated in base pairs. Lane 12 shows an assay control using 1 µg of *M. domesticus* RNA and 1 µg *M. spretus* RNA. As both protected fragments display comparable intensities the probe is equally capable of detecting both types of RNA. (Reproduced from Bartolomei et al., 1991, with permission of the publisher.)

genetic linkage more closely. No recombinants between the two genes were observed in 110 animals, which indicated that they were very close to one another (Bartolomei et al., 1991).

The absolute physical distance between the genes was established by screening a mouse yeast artificial chromosome (YAC) library (Rossi et al., 1992) for clones which contain both genes. Only one clone was obtained, and it contained both the *Igf2* gene and its nearest neighbor, the *Insulin-2* gene (Zemel et al., 1992). Further analysis of the 220 kilobase (kb) long YAC revealed that it contained sequences immediately 5′ of the *H19* gene (Fig. 5). In fact, the YAC stops just 46 bp short of the transcriptional start site of the *H19* gene. This places the 5′ ends of the *Igf2* and *H19* genes within 90 kb of DNA of each other, in the same transcriptional orientation. The gene order, in a 5′ to 3′ direction, is *Ins-2/Igf2/H19*.

There has been no similar clustering of imprinted genes reported at the only other region of the genome where an endogenous imprinted gene has been identified. Barlow and associates (1991) have shown that the *Igf2r* gene, encoding a bifunctional protein which serves as a low-affinity receptor of insulin-like growth factor II and a high-affinity receptor for mannose-6-phosphate (Kiess et al., 1988; Morgan et al., 1987), is expressed exclusively from the maternal chromosome. They have shown that *Igf2r* lies in a 800–1,100 kb segment of DNA on chromosome 17, in which at least three other genes reside. None of these is imprinted. Therefore, it remains to be seen whether the small cluster we have identified on chromosome 7 is the exception or the rule.

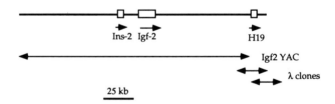

Fig. 5.   Physical map of the *Igf2-H19* locus. The physical map of 250 kb of DNA surrounding the *Igf2* and *H19* genes is indicated on the first line. The positions of the three genes are indicated by the open boxes, and the arrows below indicate their 5′ to 3′ orientation. The DNA contained in the Igf-2 YAC is indicated, with the vector arms indicated by the arrows. The physical map was extended by several λ clones containing the *H19* gene. (Reproduced from Zemel et al., 1992, and Pachnis et al., 1988, with permission of the publishers.)

VI. A Model for Dependent Imprinting of *IGF2* and *H19*

The physical linkage of *Igf2* and *H19* raised the possibility that their imprinting was coupled in some manner, although their expression from different chromosomes eliminated the most straightforward models for *cis*-acting regulators of imprinting. This possibility of a functional connection was greatly strengthened by the observation that the two genes are expressed in a very similar, if not identical pattern during development. Lee and associates (1990) and Poirier and coworkers (1991) visualized the expression of *Igf2* and *H19*, respectively, by *in situ* hybridization to mouse embryos and fetuses at various stages of development. The patterns obtained in the two independent studies were virtually identical, suggesting that the two genes operate from a common set of regulatory elements. Both genes are activated at the time of implantation in extra- embryonic tissues and by day 7.5 in the embryo proper. Expression of both genes persists in a wide array of tissues of endodermal and mesodermal origin, with no detectable expression in the central and peripheral nervous system.

One striking exception to this rule occurs in the choroid plexus and leptomeninges, two non-neural tissues in the brain. Early in development both genes are on in these two tissues, but sometime before birth, the *Ifg2* maternal allele begins to be transcribed, and both *H19* alleles shut off (Brunkow and Tilghman, 1991; DeChiara et al., 1991; J. Cunningham and S. Tilghman, unpublished observations). Thus the only tissues in which the genes are *not* imprinted are the two tissues in which they are not coexpressed.

These observations have led us to propose a model to explain the imprinting of both *Igf2* and *H19* (Bartolomei and Tilghman, 1992). It derives its inspiration from the elegant studies of J. D. Engel (Choi and Engel, 1988) and G. Felsenfeld (Nickol and Felsenfeld, 1988), who have studied the phenomenon of β-globin switching in chickens. The chicken β-globin genes lie in tandem, with the adult-specific gene 5' to the embryonic ε-globin gene. Between the genes lies a single enhancer. Early in development, the embryonic gene is exclusively transcribed, in large part because it is the inherently stronger promoter, and is able to monopolize the action of the enhancer. Later on, an adult-stage–specific transcription factor is activated, which binds to the adult β-globin promoter. As a consequence, that promoter now becomes favored in the competition for the single enhancer, and is selectively transcribed. The proof of this model was provided by Choi et al. (1988), who duplicated the enhancer. Now the two genes are transcribed simultaneously. Thus, the switch is a direct consequence of two genes competing for the action of a single enhancer.

By analogy, we propose that the imprinting of the *Igf2* and *H19* genes is mediated by their competition in *cis* for common regulatory elements. The common regulatory elements would explain their very similar patterns of expression during embryogenesis. The competition is set up by epigenetic markings on one or both of the two chromosomes, presumably placed there during gametogenesis, the only time when the chromosomes are apart and can be differentially modified. The marking(s) would act to favor transcription of *Igf2* on the paternal chromosome and *H19* on the maternal chromosome (Fig. 6). The model can accommodate either a single mark on only one chromosome, or two different marks on the two chromosomes. In the first instance, the chromosome without the mark would transcribe whichever gene is the better competitor for the regulatory elements. The mark could then act either positively or negatively to facilitate transcription of the weaker promoter on the other chromosome.

Two important requirements of this model need to be identified functionally, the regulatory elements which the genes compete for, and the mark(s)

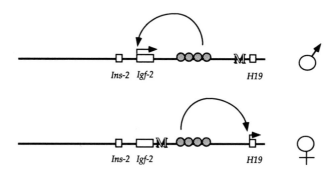

Fig. 6.   A model for dependent imprinting of *Igf2* and *H19*. The *Igf2-H19* locus is drawn as in Figure 5, with the addition of four hypothetical regulatory elements for which the *Igf2* and *H19* genes compete (mottled circles). The positions of the elements are not based on experimental evidence, and are arbitrary. **M** indicates the epigenetic mark which biases the competition for the regulatory elements. Its position is also arbitrary. On the paternal chromosome (top), the regulatory elements interact with the *Igf2* promoter to activate its transcription (indicated by the large arrow), and the *H19* gene is silent. On the maternal chromosome (bottom), the opposite occurs.

which bias the competition. We know very little about the regulatory elements that govern expression of either *Igf2* or *H19* at the moment. Only two enhancers have been identified to date; both lie 3' to the *H19* gene, and have been shown to activate its transcription in tissue culture cells derived from the liver and gut (Yoo-Warren et al., 1988), as well as in transgenic mice (Brunkow and Tilghman, 1991). These two enhancers must certainly represent but a small subset of the enhancers which govern expression of *H19*, as they are not sufficient to drive expression of the *H19* transgene in many other tissues of both endodermal and mesodermal origin. Whether these two enhancers have a role in the expression of the *Igf2* gene or not remains to be determined.

An imprinting mark must satisfy several important criteria. First, it must be established in the male and/or female germline. Second, it must be capable of stable and faithful propagation through many cell divisions, as the imprint remains in effect even in old animals (Bartolomei et al., 1991; M. Bartolomei and S.M. Tilghman, unpublished results). Finally the mark must be able to be erased and reset in the germline of the next generation.

One very strong candidate for such a mark is DNA methylation, for the practical reason that it can fulfill the three criteria. By positing oocyte- and/or spermatocyte-specific methylation patterns, the mark could be set. The action of a very efficient hemimethylase, incapable of methylating unmethylated DNA, is sufficient to maintain the mark in an allele-specific manner in somatic tissue of the next generation. Finally, a demethylase in the germline is the eraser. Studies are underway in several laboratories to ascertain whether there are chromosome-specific differences in methylation within the three known imprinted loci.

Can the competition model be generalized to other imprinted loci? The only other imprinted locus which has been studied at the molecular level is the *Tme* region on mouse chromosome 17 (Barlow et al., 1991). As indicated earlier, only one imprinted gene has been discovered in this region, and it is exclusively expressed from the maternal chromosome, akin to *H19*. It has been suggested recently, based on the survival of animals which lack *Igf2r*, that another imprinted gene resides within the *Tme* region that is responsible for the lethal effects of inheriting the deletion exclusively from mothers (Forejt and Gregorova, 1992). However, other interpretations of this result are possible and the hypothetical gene has not been identified. In humans, two syndromes with apparent "opposite" phenotypes, Prader-Willi and Angelman syndromes, map to the same region on chromosome 15q11–13.

Deletions associated with Prader-Willi show phenotypes when inherited from fathers, and Angelman deletions are deleterious when inherited from mothers (Knoll et al., 1989). For some time, these two syndromes were thought to represent reciprocal gain and loss of function of the same gene product. However, it has recently been shown that they can be genetically separated (Saitoh et al., 1992), leaving open the possibility that the syndromes represent two different, tightly linked and oppositely imprinted genes.

VII. *H19*: An RNA in Search of a Function

The foregoing model does not require that the product of the *H19* gene play any role in the imprinting itself, only that its transcriptional apparatus participate in a competition with that of the *Igf2* gene. However, the model could suggest one resolution to the mystery of the unusual nature of the *H19* gene product as a noncoding mRNA. That is, the sole function of the *H19* gene is to act as the transcriptional "foil" to facilitate the imprinting of *Igf2*. Put another way, the only important domain of the *H19* gene would be its transcriptional regulatory apparatus. This suggestion, however, is difficult to reconcile with the evolutionary conservation of both primary and secondary structures of the *H19* gene (Brannan et al., 1990; Tilghman et al., 1992). If the gene was acting only as a transcription unit, what was being transcribed would, presumably, not be conserved.

Another scenario is suggested by the similarities between *H19* and *XIST*, a gene which is expressed in an allele-specific manner on the X chromosome (Brown et al., 1991; Brockdorff et al., 1991; Pizzuti et al., 1991). The gene *XIST* maps to the X inactivation center of the mammalian X chromosome, and is exclusively expressed from the inactive X chromosome. Like *H19*, the *XIST* gene does not encode an open reading frame that is conserved between humans and mice, at least in the partial sequences published to date (Pizzuti et al., 1991). Should this hold for the entire transcript, then both *H19* and *XIST* encode RNAs with no known protein products, and are expressed in an allele-specific manner from chromosomes on which the neighboring genes are selectively silent. If both gene products are indeed RNAs, it is possible to envisage a mechanism of action in which the RNAs function in *cis* to affect either their own transcription or to suppress the transcription of neighboring gene(s). RNAs, by virtue of the fact that they are synthesized and processed in the nucleus and require no cytoplasmic event, are good candidates for molecules which could act in *cis*. Certainly the precedent of nuclear RNAs being involved at all stages of gene expression in eucaryotes, from transcription to RNA processing and transport, tends to make this possibility less fanciful.

VIII. CONCLUSIONS

The phenomenon of parental imprinting in mammals is just now being understood at the molecular level, with the identification of imprinted genes. It provides us with a model for gene expression in which the regulation is as much in *cis* as in *trans*. In addition it is relevant to one of the most fundamental problems in developmental biology: How are patterns of gene expression stably maintained through the life of an animal? What is still unresolved is the purpose of imprinting and why it evolved in mammals and not other vertebrates.

ACKNOWLEDGMENTS

I would like to thank the current members of my laboratory who contributed to the experiments and ideas discussed in this lecture. They include Drs. Marisa Bartolomei, Sharon Zemel, Karl Pfeifer, Philip Leighton, and Jennifer Cunningham. I am also indebted to Drs. Vassilis Pachnis, Cami Brannan, and Mary Brunkow, three courageous graduate students who puzzled so enthusiastically over the *H19* gene. This work was supported by grants from the National Foundation-March of Dimes and the National Institutes of Health, and by funds provided by the Howard Hughes Medical Institute.

REFERENCES

Barlow, D.P., Stoger, R., Herrmann, B.G., Saito, K., and Schweifer, N. (1991). *Nature* **349,** 84–87.

Bartolomei, M.S., and Tilghman, S.M. (1992). *Sem. Dev. Biol.* **3,** 107–117.

Bartolomei, M.S., Zemel, S., and Tilghman, S.M. (1991). *Nature* **351,** 153–155.

Beechey, C.V., Cattanach, B.M., and Searle, A.G. (1990). *Mouse Genome* 587, 64–65.

Brannan, C.I., Dees, E.C., Ingram, R.S., and Tilghman, S.M. (1990). *Mol. Cell. Biol.* **10,** 28–36.

Brockdorff, N., Ashworth, A., Kay, G.F., Cooper, P., Smith, S., McCabe, V.M., Norris, D.P., Penny, G.D., Patel, D., and Rastan, S. (1991). *Nature* **351,** 329–331.

Brown, C.J., Ballabio, A., Rupert, J.L., Lafreniere, R.G., Grompe, M., Tonlorenzi, R., and Willard, H.F. (1991). *Nature* **349,** 38–44.

Brunkow, M.E., and Tilghman, S.M. (1991). *Genes Dev.* **5,** 1092–1101.

Cattanach, B.M. (1986). *J. Embryol. Exp. Morphol.* Suppl. **97,** 137–150.

Cattanach, B.M., Beechey, C.V., Evans, E.P., and Burtenshaw, M. (1991). *Mouse Genome* **89,** 255.

Cattanach, B.M., and Kirk, M. (1985). *Nature* **315,** 496–498.

Choi, O.-R.B., and Engel, J.D. (1988). *Cell* **55,** 17–26.

DeChiara, T.M., Efstratiadis, A., and Robertson, E.J. (1990). *Nature* **345,** 78–80.

DeChiara, T.M., Robertson, E.J., and Efstratiadis, A. (1991). *Cell* **64**, 849–859.

Ferguson-Smith, A.C., Cattanach, B.M., Barton, S.C., Beechey, C.V., and Surani, M.A. (1991). *Nature* **351**, 667–670.

Forejt, J., and Gregorova, S. (1992). *Cell* **70**, 1–20.

Graham, C.F. (1974). *Biol. Rev.* **49**, 399–422.

Han, D.K., and Liau, G. (1992). *Circ. Res.* **71**, 711–719.

Kaufman, M.H. (1983) "Early Mammalian Development: Parthenogenetic Studies." Cambridge University Press, New York.

Kiess, W., Haskell, J.F., Lee, L., Greenstein, L.A., Miller, B.E., Aarons, A.L., Rechler, M.M., and Nissley, S.P. (1988). *J. Biol. Chem.* **263**, 9339–9344.

Knoll, J.H.M., Nicholls, R.D., Magenis, R.E., Graham, J.M.J., Lalande, M., and Latt, S.A. (1989). *Am. J. Med. Genet.* **32**, 285–290.

Lalley, P.A., and Chirgwin, J.M. (1984). *Cytogenet. Cell Genet.* **37**, 515.

Lee, J.E., Pintar, J., and Efstratiadis, A. (1990). *Development* **110**, 151–159.

McGrath, J., and Solter, D. (1983). *J. Exp. Zool.* **228**, 355–362.

McGrath, J., and Solter, D. (1984). *Cell* **37**, 179–183.

McGrath, J., and Solter, D. (1986). *J. Embryol. Exp. Morphol.* **97**, 277–289.

Morgan, D.O., Edman, J.C., Standring, D.N., Fried, V.A., Smith, M.C., Roth, R.A., and Rutter, W.J. (1987). *Nature* **329**, 301–307.

Nickol, J.M., and Felsenfeld, G. (1988). *Proc. Natl. Acad. Sci. U.S.A.* **85**, 2548–2552.

Pachnis, V., Belayew, A., and Tilghman, S.M. (1984). *Proc. Natl. Acad. Sci. U.S.A.* **81**, 5523–5527.

Pachnis, V., Brannan, C.I., and Tilghman, S.M. (1988). *EMBO J.* **7**, 673–681.

Pizzuti, A., Muzny, D., Lawrence, C., Willard, H.F., Avner, P., and Ballabio, A. (1991). *Nature* **351**, 325–328.

Poirier, F., Chan, C.-T.J., Timmons, P.M., Robertson, E.J., Evans, M.J., and Rigby, P.W.J. (1991). *Development,* **113**, 1105–1114.

Rossi, J.M., Burke, D.T., Leung, J.C.M., Koos, D.S., Chen, H., and Tilghman, S.M. (1992). *Proc. Nat. Acad. Sci. U.S.A.* **89**, 2456–2460.

Saitoh, S., Kubota, T., Ohta, T., Jinno, Y., Niikawa, N., Sugimoto, T., Wagstaff, J., and Lalande, M. (1992). *Lancet* **339**, 366–367.

Searle, A.G., and Beechey, C.V. (1978). *Cytogen. Cell Genet.* **20**, 282–303.

Searle, A.G., and Beechey, C.V. (1985). *In* "Aneuploidy" (V.L. Dellarco, P.E. Voytek, and A. Hollaender, ed.), pp. 363–376, Plenum Press, New York.

Searle, A.G., and Beechey, C.V. (1990). *Genet. Res.* **56**, 237–244.

Snell, G.D. (1946). *Genetics* **31**, 157–180.

Surani, M.A.H., Barton, S.C., and Norris, M.L. (1984). *Nature* **308**, 548–550.

Surani, M.A.H., Barton, S.C., and Norris, M.L. (1986). *Cell* **45**, 127–136.

Tilghman, S.M., Brunkow, M.E., Brannan, C.I., Dees, C., Bartolomei, M.S., and Phillips, K. (1992). *In* "Nuclear Processes and Oncogenes" (P.A. Sharp, ed.) pp. 188–200, Academic Press, New York.

Yoo-Warren, H., Pachnis, V., Ingram, R.S., and Tilghman, S.M. (1988). *Mol. Cell. Biol.* **8**, 4707–4715.

Zemel, S., Bartolomei, M.S., and Tilghman, S.M. (1992). *Nature Gen.* **2**, 61–65.

PRION BIOLOGY AND DISEASES

STANLEY B. PRUSINER

Departments of Neurology and of Biochemistry and Biophysics, University of California, San Francisco, California

I. Introduction

THE transmissible prion particle is composed largely, if not entirely, of an abnormal isoform of the prion protein (PrP) designated PrP^{Sc} (1). These findings argue that prion diseases should be considered pseudoinfections since the particles transmitting disease appear to be devoid of a foreign nucleic acid and thus differ from all known microorganisms as well as viruses and viroids. Because much information, especially about scrapie of rodents and Creutzfeldt-Jakob disease (CJD) in humans, has been derived using experimental techniques adapted from virology, we continue to use terms such as infection, incubation period, transmissibility, and endpoint titration in studies of prion diseases. Indeed, discoveries in prion biology are creating a new area of investigation which lies at the intersection of cell biology, genetics, and virology.

Studies on the molecular biology and chemical structure of prions may open new avenues of investigation into fundamental mechanisms of cellular regulation and homeostasis not previously appreciated (1,2). Research with rodent models of prion diseases is beginning to elucidate mechanisms responsible for CNS degeneration. Individuals at risk for familial prion diseases can often be identified decades in advance of developing CNS dysfunction (3), yet no effective therapy exists to prevent these fatal disorders. Applied research directed at detecting prions in asymptomatic cattle and sheep is urgently needed. Bovine spongiform encephalopathy (BSE) threatens the beef industry of Great Britain (4–13) and possibly other countries. More than 85,000 cattle have died of BSE to date and the production of pharmaceuticals involving cattle-derived products, which are potentially contaminated with BSE prions, is also of concern. Control of sheep scrapie in many countries is a persistent and vexing problem (14,15).

85

The Harvey Lectures, Series 87, pages 85–114
© 1993 Wiley-Liss, Inc.

TABLE I. PRION DISEASES[a]

Disease	Natural host
Scrapie	Sheep and goats
Transmissible mink encephalopathy (TME)	Mink
Chronic wasting disease (CWD)	Mule deer and elk
Bovine spongiform encephalopathy (BSE)	Cattle
Kuru	Humans (Fore)
Creutzfeldt-Jakob disease (CJD)	Humans
Gerstmann-Sträussler-Scheinker syndrome (GSS)	Humans
Fatal familial insomnia (FFI)	Humans

[a]Alternative terminologies include slow virus infections, subacute transmissible spongiform encephalopathies, and unconventional slow virus diseases (16).

Besides scrapie and BSE, six other disorders presently comprise the ensemble of prion diseases (Table I). Like BSE, both transmissible mink encephalopathy (TME) and chronic wasting disease (CWD) of captive mule deer and elk are thought to result from the ingestion of prion-infected animal products. Kuru, CJD, and Gerstmann-Sträussler-Scheinker syndrome (GSS) are all human neurodegenerative diseases that are frequently transmissible to laboratory animals (16–20).

Neither the cause of BSE, often referred to as "mad cow disease", nor methods of controlling the spread of this disorder are known. Many investigators contend that BSE resulted from the feeding of dietary protein supplements derived from rendered scrapie-infected sheep offal to cattle, a practice banned since 1988. Curiously, the majority of BSE cases have occurred in herds with a single affected animal within a herd; several cases of BSE in a single herd are infrequent (5,7,11). Whether the distribution of BSE cases within herds will change as the epidemic progresses and whether BSE will disappear with the cessation of feeding rendered meat and bone meal are uncertain.

Of particular importance to the BSE epidemic is kuru of humans, confined to the Fore region of New Guinea (16,17). Once the most common cause of death among women and children, kuru has almost disappeared with the cessation of ritualistic cannibalism (21). These findings argue that kuru was transmitted orally as proposed for BSE. Of note are recent cases of kuru which have occurred in people exposed to prions more than three decades ago.

II. DEVELOPMENT OF THE PRION HYPOTHESIS

The unusual biological properties of the scrapie agent were first recognized in studies with sheep (22). The experimental transmission of scrapie to mice (23) gave investigators a more convenient laboratory model which yielded considerable information on the novelty of the infectious pathogen causing scrapie (24–32). Yet progress was slow, since quantitation of infectivity in a single sample required holding 60 mice for one year prior to scoring an endpoint titration (23).

Development of a more rapid and economical bioassay for the scrapie agent in Syrian golden hamsters accelerated work aimed at purification of the infectious particles (33,34). Partial purification led to the discovery that a protein is required for infectivity (35), a finding that was in agreement with some earlier studies which raised the possibility that protein might be necessary (36–38). Procedures which modify nucleic acids were found not to alter scrapie infectivity (2). Other investigators had demonstrated the extreme resistance of scrapie infectivity to both ultraviolet and ionizing radiation (24–28), prompting speculation that the scrapie pathogen might be devoid of nucleic acid—a postulate dismissed by most scientists. These early radiobiological results were extended using reagents specifically modifying or damaging nucleic acids: nucleases, psoralens, hydroxylamine, and Zn^{2+} ions— none of which were found to alter scrapie infectivity in homogenates (2), microsomal fractions (2), purified prion rod preparations, or detergent-lipid-protein complexes (DLPCs) (39–43).

Because of the foregoing results, the term ''prion'' was introduced to distinguish the *pro*teinaceous *in*fectious particles that cause scrapie, CJD, GSS, and kuru, from both viroids and viruses (2). Hypotheses for the structure of the infectious prion particle included:

1. Proteins surrounding a nucleic acid which encodes them (a virus).
2. Proteins associated with a small polynucleotide.
3. Proteins devoid of nucleic acid.

Mechanisms postulated for the replication of infectious prion particles ranged from those used by viruses to the synthesis of polypeptides in the absence of nucleic acid template, to post-translational modifications of cellular proteins. Subsequent discoveries have narrowed hypotheses for both prion structure and the mechanism of replication.

III. DISCOVERY OF THE PRION PROTEIN

Progress in the study of prions and the CNS degenerative diseases that they cause was dramatically accelerated by the discovery of a protein designated *pri*on *p*rotein or PrP (44–46). Enriching fractions from hamster brain for scrapie infectivity led to the identification of a protease-resistant protein of M_r 27–30 kd, designated PrP 27-30, that was present in scrapie fractions but absent from controls. Purification of PrP 27- 30 to homogeneity allowed determination of its N-terminal amino acid sequence (47) which, in turn, permitted the synthesis of isocoding mixtures of oligonucleotides that were used to identify cDNA clones encoding PrP (48–51). Subsequent studies revealed that PrP is encoded by a chromosomal gene and not by a nucleic acid within the infectious scrapie prion particle (49). Levels of PrP mRNA remain unchanged throughout the course of scrapie infection—an observation which led to the identification of the normal PrP gene product, a protein of 33–35 kd, designated PrP^C (48). PrP^C is protease-sensitive, while PrP 27-30 was found to be the protease-resistant core of a 33–35 kd disease-specific protein, designated PrP^{Sc} (Table II).

Molecular clones were recovered from cDNA libraries constructed from mRNA isolated from scrapie-infected Syrian hamster (SHa) and mouse (Mo) brains. The translated sequences of the SHa and Mo PrP cognate cDNAs encode proteins of 254 amino acids (Fig. 1) (48–51). Identical sequences were deduced from genomic clones derived from DNA isolated from uninfected, control animals (49). Human (Hu) PrP contains 253 residues (52). Signal peptides of 22 amino acids at the N-terminus are cleaved during the biosynthesis of SHa and Mo PrP in the rough endoplasmic reticulum (53–55). A 23-residue peptide is removed from the C-terminus of SHaPrP upon addition of a glycoinositol phospholipid (GPI) anchor (56–59). Two Asn-linked complex-type oligosaccharides are attached to sites within a loop formed by a disulfide bond (54,60–64). Limited proteolysis of PrP^{Sc} generates PrP 27-30 which lacks ~67 amino acids from the N-terminus of PrP^{Sc} (47,48). Neither gas-phase sequencing nor mass spectrometric analysis of PrP 27-30 has revealed any amino acid differences between the sequence determined by these methods and that deduced from the translated sequence of molecular clones (65). Conclusions about the covalent structure of PrP^{Sc} must be guarded, since purified fractions contain ~10^5 PrP 27-30 molecules/ID_{50} unit (45). If <1% of the PrP 27- 30 molecules contained an amino acid substitution or post-translational chemical modification which conferred scrapie infectivity, our methods would not detect such a change.

IV. Infectious Prion Particles

A remarkable convergence of information on PrP^{Sc} in prion diseases argues persuasively that prions are composed largely, if not entirely, of PrP^{Sc} molecules (Table III). Although some investigators contend that PrP^{Sc} is

TABLE II. PROPERTIES OF CELLULAR AND SCRAPIE PrP ISOFORMS

Property	PrP^C	PrP^{Sc}	References
Concentration in normal SHa brain	~1–5 µg/g	—	(180)
Concentration in scrapie-infected SHa brain	~1–5 µg/g	~5–10 µg/g	(180)
Presence in purified prions	—	+ [a]	(44–47,88)
Protease resistance	—	+ [b]	(44–48,88)
Presence in amyloid rods	—	+ [c]	(34,36,88–93,103)
Subcellular localization in cultured cells	Cell surface	Primarily cytoplasmic vesicles	(56,125,126)
PIPLC[d] release from membranes	+	—	(56,124)
Synthesis ($t_{1/2}$)	<0.1 h	~1–3 h [e]	(121,122, 200)
Degradation ($t_{1/2}$)	~5 h	>>24 h	(121,122, 200)

[a]Copurification of PrP^{Sc} and prion infectivity demonstrated by two protocols: 1) detergent extraction followed by sedimentation protease digestion, and 2) PrP 27–30 monoclonal antibody affinity chromatography.

[b]Limited proteinase K digestion of SHaPrP^{Sc} produces PrP 27–30.

[c]After limited proteolysis of PrP^{Sc} (PrP 27–30 is produced) and detergent extraction, amyloid rods form; except for length, the rods are indistinguishable from amyloid filaments forming plaques.

[d]PIPLC, phosphatidylinositol-specific phospholipase C.

[e]PrP^{Sc} de novo synthesis is a post-translational process.

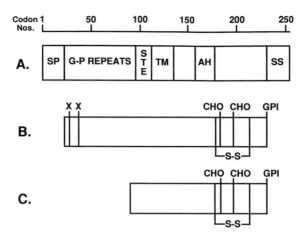

Fig. 1. Structural features of the Syrian golden hamster prion protein. Codon numbers are indicated at the top of the figure. **A:** NH$_2$- terminal signal peptide (SP) of 22 amino acids is removed during biosynthesis (48,49,53–55). The NH$_2$-terminal region contains five Gly-Pro-rich (G-P; rich in glycine and proline) octarepeats and two hexarepeats; between codons 96 and 112 a domain controlling PrP topology is designated as the stop-transfer effector (STE) (129,130); codons 113 to 135 encode a transmembrane (TM) α-helix; codons 157 to 177 encode an amphipathic helix (AH) (129–133); and codons 232 to 254 encode a hydrophobic signal sequence (SS) which is removed when a GPI anchor is added (56,57). **B:** Unknown modifications (X) of the arginine residue at codons 25 and 37 in PrP^{Sc} and at least codon 25 in PrP^C result in a loss of the arginine signal in the Edman degradation, but these modifications are inconsistently reported (54). Both PrP isoforms contain a disulfide (S–S) bond between Cys^{179} and Cys^{214} (54); asparagine-linked glycosylation (CHO) occurs at residues 181 and 197 (60–64), and a GPI anchor is attached to Ser^{231} (57). **C:** PrP 27-30. This molecule is derived from PrP^{Sc} by limited proteolysis that removes the 67 NH$_2$-terminal amino acids and leaves a protease-resistant core of 141 amino acids (48,49).

merely a pathologic product of scrapie infection and that PrP^{Sc} coincidentally purifies with the ''scrapie virus'' (66–73), no convincing data to support this view have been offered. No fractions containing <1 PrP^{Sc} molecule per ID$_{50}$ unit have been found; such fractions would argue that PrP^{Sc} is not required for infectivity. Some investigators have reported that PrP^{Sc} accumulation in hamsters occurs after the synthesis of infectivity (74,75), but these results have been refuted by three groups working independently (76,77,201). The discrepancy appears to be due to comparisons of infectivity in crude homogenates, with PrP^{Sc} concentrations measured in purified fractions.

TABLE III. EVIDENCE THAT PrP^{Sc} IS A MAJOR AND NECESSARY
COMPONENT OF THE INFECTIOUS PRION

1. Copurification of PrP 27–30 and scrapie infectivity by biochemical methods. Concentration of PrP 27–30 is proportional to prion titer (44–46).
2. Kinetics of proteolytic digestion of PrP 27–30 and infectivity are similar (46).
3. Copurification of PrP^{Sc} and infectivity by immunoaffinity chromatography, α-PrP antisera neutralization of infectivity (104,105).
4. PrP^{Sc} detected only in clones of cultured cells producing prion infectivity (125,127).
5. PrP amyloid plaques are specific for prion diseases of animals and humans (89–92). Deposition of PrP amyloid is controlled, at least in part, by the PrP sequence (174).
6. Correlation between PrP^{Sc} (or PrP^{CJD}) in brain tissue with prion diseases in animals and humans (190–192).
7. Genetic linkage between MoPrP gene and scrapie incubation times (138–141). PrP gene of mice with long incubation times encodes amino acid substitutions at codons 108 and 189 as compared to mice with short or intermediate incubation times (109).
8. SHaPrP transgene and scrapie PrP^{Sc} in the inoculum govern the "species barrier," scrapie incubation times, neuropathology, and prion synthesis in mice (174,183).
9. Genetic linkage between human PrP gene mutations at codons 102, 198, or 200 and development of GSS or familial CJD (3) as well as linkage between insertion of six additional octarepeats and familial CJD (145–149,157,193).
10. Mice expressing MoPrP transgenes with the point mutation of GSS spontaneously develop neurologic dysfunction, spongiform brain degeneration, and astrocytic gliosis (175).

The search for a second component within the prion particle has focused largely on a nucleic acid because it would most readily explain different isolates or "strains" of infectivity (78–81). Multiple scrapie isolates, each with different incubation times, have been found to breed true in mice and hamsters (78–81). In addition, many other factors modulate scrapie incubation times including: PrP gene expression, murine genes linked to PrP designated *Prn-i* and *Sinc*, dose of inoculum, route of inoculation, and the genetic origin of the prion inoculum, that is, PrP^{Sc} sequence.

The search for a scrapie-specific nucleic acid molecule has been unrewarding despite the use of reagents that modify or hydrolyze polynucleotides, molecular cloning procedures, and physicochemical techniques (39–43, 82–87). The implications of an infectious pathogen that does not contain a nucleic acid are profound. Although available data does not permit its exclusion, finding a scrapie-specific polynucleotide seems unlikely. The possibility of a noncovalently bound cofactor such as a peptide, oligosaccharide, fatty acid, sterol, or inorganic compound also deserves consideration (see Fig. 4).

V. PrP Amyloid

The discovery of PrP 27-30 in fractions enriched for scrapie infectivity was accompanied by the electron microscopic identification of rod-shaped particles in rapidly sedimenting fractions of discontinuous sucrose gradients (44–46,88). The rods are ultrastructurally indistinguishable from many purified amyloids and display the tinctorial properties of amyloids (88). These findings were followed by the demonstration that amyloid plaques in prion diseases contain PrP, as determined by immunoreactivity and amino acid sequencing (89–93). It has been claimed that scrapie-associated fibrils (SAF) are synonymous with prion rods and are composed of PrP, although SAF was repeatedly distinguished from amyloids (94–102).

Studies on the generation of prion rods have shown that their formation requires limited proteolysis in the presence of detergent (103). Thus, the prion rods found in fractions enriched for scrapie infectivity are largely, if not entirely, artifacts of the purification protocol. While the prion rods gave the first clue that amyloid plaques in prion diseases might be composed of PrP (88), these insoluble structures have greatly retarded protein chemical analysis of PrP^{Sc}. Functional solubilization of PrP 27-30 in DLPCs with retention of infectivity (87) demonstrated that PrP polymers are not required for infectivity and permitted the immunoaffinity copurification of PrP^{Sc} and infectivity (104,105).

VI. PrP Gene Structure and Expression

Mapping PrP genes to the short arm of human chromosome 20 and the homologous region of mouse chromosome 2 argues for the existence of PrP genes prior to the speciation of mammals (106–108). Hybridization studies demonstrated <0.002 PrP gene sequences per ID$_{50}$ unit in purified prion fractions indicating that a gene encoding PrP^{Sc} is not a component of the infectious prion particle (48). This is a major feature which distinguishes prions from viruses including those retroviruses that carry cellular oncogenes, and from satellite viruses that derive their coat proteins from other viruses previously infecting plant cells.

Since a single exon contains the entire open reading frame (ORF) of all known PrP genes, the possibility that variant forms of PrP arise from alternative RNA splicing is unlikely (49,109,110). Although PrP^{Sc} sequencing studies, as noted above, argue that RNA editing or protein splicing (111,112) are unlikely mechanisms for generating the PrP^{Sc} isoform, they cannot be formally excluded. The two exons of the SHaPrP gene are separated by a 10 kb intron: exon 1 encodes a portion of the 5′ untranslated leader sequence, while exon 2 encodes the ORF and 3′ untranslated region (49). The MoPrP gene is comprised of three exons with exon 3 analogous to exon 2 of the hamster (110). The promoters of both the SHa and MoPrP genes contain copies of G-C–rich repeats 3 and 2, respectively, but are devoid of TATA boxes. These G-C (guanine-cytosine) nonamers represent a motif which may function as a canonical binding site for the transcription factor Sp1 (113).

Although PrP mRNA is constitutively expressed in the brains of adult animals (48), it is highly regulated during development. In the septum, levels of PrP mRNA and choline acetyltransferase were found to increase in parallel during development (114). In other brain regions, PrP gene expression occurred at an earlier age. In situ hybridization studies show that the highest levels of PrP mRNA are found in neurons (115).

Four regions of the mammalian PrP gene ORF are highly conserved when the translated amino acid sequences are compared (Fig. 2) (48–52,116–118) (M. Scott et al., unpublished data). While the function of PrP^{C} is unknown, the MoPrP sequence is ~30% identical with a putative factor from chickens possessing acetylcholine-receptor–inducing activity (ARIA) (119,120). Twenty-three of 24 amino acids encoded by mouse *Prn-p^{a}* correspond to codons 104 to 127 and are identical to those found in ARIA. Within the N-terminal conserved regions of mammalian PrP, five octarepeats that are rich in Gly and Pro residues as well as two hexarepeats (Fig. 1) have been found, while chicken ARIA has eight hexarepeats.

VII. SYNTHESIS OF PRP ISOFORMS

Metabolic labeling studies of scrapie-infected cultured cells have shown that PrP^{C} is synthesized and degraded rapidly while PrP^{Sc} accumulates slowly (121–123) (D. R. Borchelt et al., in preparation). These observations are consistent with earlier findings showing that PrP^{Sc} accumulates to high levels in the brains of scrapie-infected animals, yet PrP mRNA levels remain unchanged (48).

Both PrP isoforms appear to transit through the Golgi apparatus where their Asn-linked oligosaccharides are modified and sialylated (60–64). PrP^{C}

is presumably transported within secretory vesicles to the external cell surface where it is anchored by a gycosylinositol phospholipid (GPI) moiety (56–59, 124). In contrast, PrP^{Sc} accumulates primarily within cells, where it is deposited in cytoplasmic vesicles, many of which appear to be secondary lysosomes (Table II) (125–127). Although most of the difference in mass of PrP 27-30 predicted from the amino acid sequence and that observed after post-translational modification is due to complex-type oligosaccharides, these sugar chains are not required for the synthesis of protease-resistant PrP in scrapie-infected cultured cells; this conclusion is based on experiments with the Asn-linked glycosylation inhibitor tunicamycin and on site-directed mutagenesis studies (128). Whether unglycosylated PrP^{Sc} is associated with scrapie prion infectivity remains to be established, but experiments with transgenic mice may resolve this issue.

Cell-free translation studies have demonstrated two forms of PrP: a transmembrane form which spans the bilayer twice at the transmembrane (TM) and amphipathic helix (AH) domains, and a secretory form (Fig. 1) (129–133). The stop-transfer effector (STE) domain controls the topogenesis of PrP. That PrP contains both a TM domain and a GPI anchor poses a topologic conundrum. It seems likely that membrane-dependent events feature in the synthesis of PrP^{Sc}, especially since brefeldin A, which selectively destroys the Golgi stacks (134,135), prevents PrP^{Sc} synthesis in scrapie-infected cultured cells (136). For many years, the association of scrapie infectivity with membrane fractions has been appreciated (29–31); indeed, hydrophobic interactions are thought to account for many of the physical properties displayed by infectious prion particles (33,87,137).

VIII. GENETIC LINKAGE OF PrP WITH SCRAPIE INCUBATION TIMES

Studies of PrP genes (*Prn-p*) in mice with short and long incubation times demonstrated genetic linkage between a *Prn-p* restriction fragment length polymorphism (RFLP) and a gene modulating incubation times (*Prn-i*) (138). Other investigators have confirmed the genetic linkage, and one group has shown that the incubation time gene *Sinc* is also linked to PrP (139–141). The gene *Sinc* was first described by Dickinson and colleagues over 20 years ago (142); whether the genes for PrP, *Prn-i*, and *Sinc* are all congruent remains to be established. The PrP sequences of NZW (*Prn-p^a*) and I/Ln (*Prn-p^b*) mice with short and long scrapie incubation times, respectively, differ at codons 108 (L→F) and 189 (T→V) (Fig. 2) (109). While these amino acid substitutions argue for the congruency of *Prn-p* and *Prn-i*, experiments with *Prn-p^a* mice expressing *Prn-p^b* transgenes demonstrated a paradoxical short-

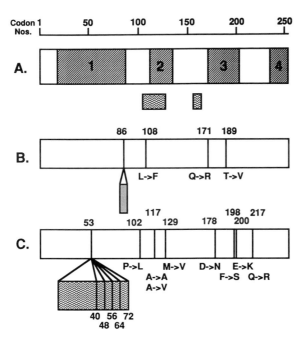

Fig. 2. Genetic map of prion protein open reading frames. Codon numbers are indicated at the top of the figure. **A:** Four regions among mammalian PrP molecules (hatched) (48–52,116–118,202); regions of MoPrP homologous to a molecule found in fractions containing acetylcholine receptor-inducing activity in chickens (wave) (120,202). **B:** Animal mutations and polymorphisms. Two alleles of bovine PrP have been identified, with one containing an additional octarepeat (stippled) at codon 86, and a polymorphism at codon 171 in sheep PrP resulting in the substitution of arginine for glutamine (117,194,202). Mice with $Prn-p^b$ genes have long scrapie incubation times, and amino acid substitutions at codon 108 (Leu→Phe) and 189 (Thr→Val) (109). **C:** Human PrP mutations and polymorphisms. Octarepeat inserts of 32, 40, 48, 56, 64, and 72 amino acids have been found (145–149,152,195). Inserts of 40, 48, 56, 64, and 72 amino acids are associated with familial CJD. Point mutations at codons 102 (Pro→Leu), 117 (Ala→Val), and 198 (Phe→Ser) are found in patients with GSS (3,156,160,161,178,195–199). There are common polymorphisms at codons 117 (Ala→Ala) and 129 (Met→Val). Point muations at codons 178 (Asp→Asn) and 200 (Glu→Lys) are found in patients with familial CJD (157,158,162,193). Point mutations at codons 198 (Phe→Ser) and 217 (Gln→Arg) are found in patients with GSS who have PrP amyloid plaques and neurofibrillary tangles (K. Hsiao et al., submitted for publication). Single letter code for amino acids is as follows: A, Ala; D, Asp; E, Glu; F, Phe; K, Lys; L, Leu; M, Met; N, Asn; P, Pro; Q, Gln; R, Arg; S, Ser; T, Thr; and V, Val.

ening of incubation times (110) instead of a prolongation as predicted from $(Prn\text{-}p^a \times Prn\text{-}p^b)$ F1 mice, which exhibit long incubation times that are dominant (138–142). Whether this paradoxical shortening of scrapie incubation times in Tg($Prn\text{-}p^b$) mice results from high levels of PrP^C-B expression remains to be established (110). (PrP^C-B is the cellular isoform of the prion protein encoded by the mouse $Prn\text{-}p^b$ gene.)

Host genes also influence the development of scrapie in sheep. Parry argued that natural scrapie is a genetic disease which could be eradicated by proper breeding protocols (14,15). He considered its transmission by inoculation of importance primarily for laboratory studies and that communicable infection was of little consequence in nature. Other investigators viewed natural scrapie as an infectious disease and argued that host genetics only modulates susceptibility to an endemic infectious agent (143). The incubation time gene for experimental scrapie in Cheviot sheep called Sip is said to be linked to a PrP gene RFLP (144), a situation perhaps analogous to $Prn\text{-}i$ and $Sinc$ in mice. However, the null hypothesis of nonlinkage has yet to be tested and this is important, especially in view of earlier studies which argue that susceptibility of sheep to scrapie is governed by a recessive gene (14,15). In a Suffolk sheep, a polymorphism in the PrP ORF was found at codon 171 (R→Q) (Fig. 2B) (116,117) (D. Westaway and S.B. Prusiner, unpublished data); whether it segregates with a Sip phenotype in Cheviot sheep is unknown.

IX. INHERITED HUMAN PRION DISEASES

Genetics were first thought to have a role in CJD of humans when it was recognized that ~10% of cases are familial (16,19). Like sheep scrapie, the relative contributions of genetic and infectious etiologies in the human prion diseases remained puzzling. The discovery of the PrP gene raised the possibility that mutation might feature in the hereditary human prion diseases. A point mutation at codon 102 (P→L) was shown to be linked to development of GSS (Fig. 2C) (3). This mutation may be due to the deamination of a methylated CpG in a germline PrP gene, with the result that T is substituted for C.

An insert of 144 bp at codon 53, containing six octarepeats, has been described in patients with CJD from four families all residing in southern England (Fig. 2C) (145–150). This mutation must have arisen through a complex series of events since the HuPrP gene contains only five octarepeats which indicate that a single recombination event could not have created the insert. Genealogic investigations have shown that all four families are related, arguing for a single founder born more than two centuries ago (149).

Five, six, seven, eight, or nine octarepeats (in addition to the normal five) were found in individuals with CJD, whereas deletion of one octarepeat or four additional octarepeats have been identified without the neurologic disease (145–148,151,152).

For many years the unusually high incidence of CJD among Israeli Jews of Libyan origin was thought to be due to the consumption of lightly cooked sheep brain or eyeballs (153–155). Recent studies have shown that some Libyan and Tunisian Jews in families with CJD have a PrP gene point mutation at codon 200 resulting in a E→K substitution (156–158). One patient was homozygous for the mutation, but her clinical presentation was similar to that of heterozygotes (158) arguing that familial prion diseases are true autosomal dominant disorders like Huntington's disease (159). The codon 200 mutation has also been found in Slovaks originating from Orava in North Central Czechoslovakia (156).

Other point mutations at codons 117, 178, 198, and 217 also segregate with inherited prion diseases (160–163). Some patients once thought to have familial Alzheimer's disease are now known to have prion diseases on the basis of PrP immunostaining of amyloid plaques and PrP gene mutations (164–167). Patients with the codon 198 mutation have numerous neurofibrillary tangles that stain with antibodies to tau (τ) protein and have amyloid plaques (164–167) that are composed largely of a PrP fragment extending from residues 58 to 150 (93).

At PrP codon 129, an amino acid (Met/Val) polymorphism (Fig. 2) has been identified (168). Following treatment with human pituitary growth hormone (169, 170) or gonadotrophin, patients with CJD have a significant preponderance of the Val allele (171) compared to the general population. Sporadic CJD patients were found to be homozygous for the Met or Val allele at codon 129 but were rarely heterozygous (172). This finding was interpreted (172,173) as consistent with the hypothesis that PrP^C/PrP^{Sc} heterodimers feature in the replication of prions (1,174).

X. De Novo Synthesis of Prions in Transgenic Mice Expressing GSS Mutant MoPrP

When the codon 102 point mutation was introduced into MoPrP in transgenic mice, spontaneous CNS degeneration occurred. This was characterized by clinical signs indistinguishable from experimental murine scrapie and neuropathology and consisted of widespread spongiform morphology and astrocytic gliosis (175). By inference, these results suggest that PrP mutations cause GSS and familial CJD. It is unclear whether low levels of protease-resistant PrP in the brains of transgenic mice with the GSS mutation is PrP^{Sc} or residual PrP^{C}. Undetectable or low levels of PrP^{Sc} in the brains of these transgenic mice are consistent with the results of transmission experiments that suggest low titers of infectious prions. Brain extracts transmit CNS degeneration to inoculated recipients and the *de novo* synthesis of prions has been demonstrated by serial passage (176). These observations indicate that prions are devoid of foreign nucleic acid, in accord with studies that use other experimental approaches (39–43,82–86,177).

One view of the PrP gene mutations has been that they render individuals susceptible to a common "virus" (101,102). In this scenario, the putative scrapie virus is thought to persist within a worldwide reservoir of humans, animals, or insects without causing detectable illness. Yet one in a million individuals develops sporadic CJD and dies from a lethal "infection," while almost 100% of people with PrP point mutations or inserts appear to eventually develop neurologic dysfunction. That germline mutations found in the PrP genes of patients and at-risk individuals are the cause of familial prion diseases is supported by experiments with Tg(GSS MoPrP) mice described above (176,178,179). The transgenic mouse studies also argue that sporadic CJD might arise from the spontaneous conversion of PrP^{C} to PrP^{CJD}, due to either a somatic mutation of the PrP gene or a rare event involving modification of wild-type PrP^{C} (1).

XI. Transgenetics and Species Barriers

Passage of prions between species is a stochastic process characterized by prolonged incubation times (32,180,181). Prions synthesized de novo reflect the sequence of the host PrP gene and not that of the PrP^{Sc} molecules in the inoculum (182). On subsequent passage in a homologous host, the incubation time shortens to that recorded for all subsequent passages and it becomes a nonstochastic process. The species barrier concept is that of practical importance in assessing the risk for humans of developing CJD after consumption of scrapie-infected lamb or BSE beef.

To test the hypothesis that differences in PrP gene sequences might be responsible for the species barrier, transgenic mice expressing SHaPrP were constructed and bred (174,183). The PrP genes of Syrian hamsters and mice encode proteins differing at 16 positions. Incubation times in four lines of Tg(SHaPrP) mice inoculated with Mo prions were prolonged compared to those observed for nontransgenic, control mice (Fig. 3A). Inoculation of

Tg(SHaPrP) mice with SHa prions demonstrated abrogation of the species barrier, resulting in abbreviated incubation times due to a nonstochastic process (Fig. 3B) (174,183). The length of the incubation time after inoculation with SHa prions was inversely proportional to the level of SHaPrP^C in the brains of Tg(SHaPrP) mice (Figs. 3B and 3C) (174). SHaPrP^{Sc} levels in the brains of clinically ill mice were similar in all four Tg(SHaPrP) lines inoculated with SHa prions (Fig. 3D). Bioassays of brain extracts from clinically ill Tg(SHaPrP) mice inoculated with Mo prions revealed that only Mo prions but no SHa prions were produced (Fig. 3E). Conversely, inoculation of Tg(SHaPrP) mice with SHa prions led to only the synthesis of SHa prions (Fig. 3F). Thus, the de novo synthesis of prions is species specific and reflects the genetic origin of the inoculated prions. Similarly, the neuropathology of Tg(SHaPrP) mice is determined by the genetic origin of the prion inoculum. Mo prions injected into Tg(SHaPrP) mice produced a neuropathology characteristic of mice with scrapie. A moderate degree of vacuolation in both the gray and white matter was found, while amyloid plaques were rarely detected (Fig. 3G). Inoculation of Tg(SHaPrP) mice with SHa prions produced intense vacuolation of the gray matter, sparing of the white matter, and numerous SHaPrP amyloid plaques characteristic of Syrian hamsters with scrapie (Fig. 3H).

These studies with transgenic mice establish that the PrP gene influences virtually all phases of scrapie including:

1. Species barrier.
2. Replication of prions.
3. Incubation times.
4. Synthesis of PrP^{Sc}.
5. Neuropathologic changes.

XII. Prion Replication

Although the search for a scrapie-specific nucleic acid continues to be unrewarding, some investigators steadfastly cling to the notion that this putative polynucleotide drives prion replication. If prions are found to contain a scrapie-specific nucleic acid, then such a molecule would be expected to direct scrapie agent replication using a strategy similar to that employed by viruses (Fig. 4A). In the absence of any chemical or physical evidence for a scrapie-specific polynucleotide (Fig. 3A) (39–43,66–73,82–86,177), it seems reasonable to consider some alternative mechanisms that might feature in prion biosynthesis. The multiplication of prion infectivity is an exponential process in which the post-translational conversion of PrP^C or a precursor to

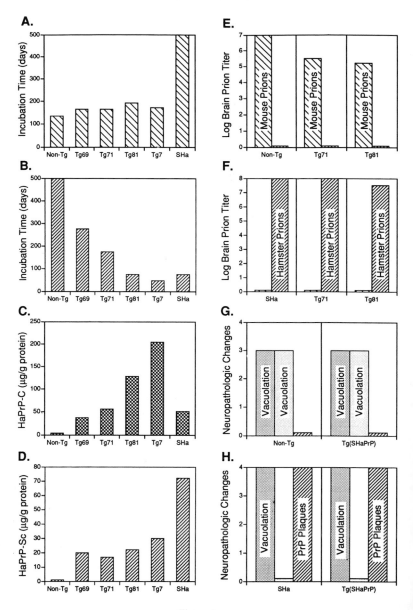

Figure 3.

PrP^{Sc} appears to be obligatory (121). As illustrated in Figure 4B, two PrP^{Sc} molecules combine with two heterodimers which are subsequently transformed into two homodimers. In the next cycle, four PrP^{Sc} molecules combine with four PrP^C molecules giving rise to four homodimers that dissociate to combine with eight PrP^C molecules creating an exponential process. Studies with Tg(SHaPrP) mice argue that prion synthesis involves "replication", not merely "amplification" (174). Assuming prion biosynthesis simply involves amplification of post-translationally altered PrP molecules, we might expect Tg(SHaPrP) mice to produce both SHa and Mo prions after inoculation with either prion since these mice produce both SHa and MoPrP^C. Yet Tg(SHaPrP) mice synthesize only those prions present in the inoculum (Figs. 3E and 3F). These results argue that the incoming prion and PrP^{Sc} interact with the homologous PrP^C substrate to replicate more of the same prions (Fig. 4C).

In the absence of any candidate post-translational chemical modifications (203) that differentiate PrP^C from PrP^{Sc}, we are forced to consider the

---

Fig. 3.   Transgenic mice expressing SHa prion protein exhibit species- specific scrapie incubation times, infectious prion synthesis, and neuropathology (174). **A:** Scrapie incubation times in nontransgenic mice (Non- Tg) and four lines of transgenic mice expressing SHaPrP and Syrian hamsters inoculated intracerebrally with $\sim 10^6$ ID$_{50}$ units of Chandler Mo prions serially passaged in Swiss mice. The four lines of transgenic mice have different numbers of transgene copies: Tg69 and 71 mice have two to four copies of the SHaPrP transgene, whereas Tg81 mice have 30 to 50 and Tg7 mice have >60. Incubation times are the number of days from inoculation to onset of neurologic dysfunction. **B:** Scrapie incubation times in mice and hamsters inoculated with $\sim 10^7$ ID$_{50}$ units of Sc237 prions serially passaged in Syrian hamsters and as described in (A). **C:** Brain SHaPrP^C in transgenic mice and hamsters. SHaPrP^C levels were quantitated by an enzyme-linked immunoassay. **D:** Brain SHaPrP^{Sc} in transgenic mice and hamsters. Animals were killed after exhibiting clinical signs of scrapie. SHaPrP^{Sc} levels were determined by immunoassay. **E:** Prion titers in brains of clinically ill animals after inoculation with Mo prions. Brain extracts from Non-Tg, Tg71, and Tg81 mice were bioassayed for prions in mice (left) and hamsters (right). **F:** Prion titers in brains of clinically ill animals after inoculation with SHa prions. Brain extracts from Syrian hamsters as well as Tg71 and Tg81 mice were bioassayed for prions in mice (left) and hamsters (right). **G:** Neuropathology in Non-Tg mice and Tg(SHaPrP) mice showing clinical signs of scrapie after inoculation with Mo prions. Vacuolation in grey (left) and white matter (center); PrP amyloid plaques (right). Vacuolation score: 0 = none, 1 = rare, 2 = modest, 3 = moderate, 4 = intense. **H:** Neuropathology in Syrian hamsters and transgenic mice inoculated with SHa prions. Degree of vacuolation and frequency of PrP amyloid plaques as described in (G).

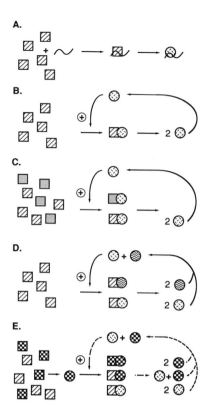

Fig. 4. Some possible mechanisms of prion replication. **A:** Two-component prion model. Prions contain a putative, as yet unidentified, nucleic acid, or another second component (solid, thick wavy line) that binds to PrP^C (squares) and stimulates conversion of PrP^C or a precursor to PrP^{Sc} (circles). **B:** One-component prion model— prions devoid of nucleic acid. PrP^{Sc} binds to PrP^C forming heterodimers that function as replication intermediates in the synthesis of PrP^{Sc}. Repeated cycles of this process result in an exponential increase in PrP^{Sc}. **C:** Prion synthesis in transgenic mice (174). SHaPrP^{Sc} (broken, cross-hatched circles) binds to SHaPrP^C (diagonal squares), leading to the synthesis of PrP^{Sc}. Binding to MoPrP^C (stippled squares) does not produce PrP^{Sc}. Species barrier for scrapie between mice and hamsters is represented by the MoPrP^C-SHaPrP^{Sc} heterodimer. **D:** Scrapie isolates or strains in hamsters or mice. Multiple PrP^{Sc} conformers (cross-hatched and wavy patterns in circles) bind to PrP^C and constrain the conformational changes that PrP^c undergoes during its conversion into PrP^{Sc}. **E:** Inherited prion diseases in humans and transgenic mice. Mutant PrP^C molecules (checkered pattern in squares) might initiate the conversion of PrP^C to

possibility that conformation distinguishes these isoforms. Various "strains" or isolates of scrapie prions (78–81) could be accommodated by multiple conformers that act as templates for the folding of *de novo* synthesized PrP^{Sc} molecules during prion "replication" (Fig. 4D). Although this proposal is rather unorthodox, it is consistent with observations generated from Tg(SHaPrP)Mo studies which contend that PrP^{Sc} in the inoculum binds to homologous PrP^C or a precursor to form a heterodimeric intermediate in the replication process (174). Whether foldases, chaperonins, or other types of molecules feature in the conversion of the PrP^C/PrP^{Sc} heterodimer to a PrP^{Sc} homodimer is unknown. The molecular weight of a PrP^{Sc} homodimer is consistent with the ionizing radiation target size of 55,000 $\pm$ 9,000 daltons, as determined for infectious prion particles independent of their polymeric form (184).

In humans carrying point mutations or inserts in their PrP genes, mutant PrP^C molecules might spontaneously convert into PrP^{Sc} (Fig. 4E). While the initial stochastic event may be inefficient, once it happens, the process becomes autocatalytic. The proposed mechanism is consistent with individuals harboring germline mutations who do not develop CNS dysfunction for decades, and with studies on Tg(GSS MoPrP) mice that spontaneously develop CNS degeneration (175). Whether all GSS and familial CJD cases contain infectious prions, or whether some represent inborn errors of PrP metabolism in which neither PrP^{Sc} nor prion infectivity accumulates, is unknown. If the latter is found, then presumably, mutant PrP^C molecules alone can produce CNS degeneration.

Some investigators have suggested that scrapie agent mutliplication proceeds through a crystallization process involving PrP amyloid formation (185–187). Against this hypothesis is the absence or rarity of amyloid plaques in many prion diseases, as well as the inability to identify any amyloid-like polymers in cultured cells that chronically synthesize prions (103,174). Purified infectious preparations isolated from scrapie-infected hamster brains exist as amorphous aggregates; only if PrP^{Sc} is exposed to detergents and limited proteolysis, does it polymerize into prion rods exhibiting the ultra-

PrP^{Sc} (or PrP^{CJD}). If infectious prions are produced (dashed lines), then they stimulate the synthesis of more PrP^{CJD} in humans and PrP^{Sc} in experimental animals. Alternatively, prion infectivity is not generated, but the host develops neurologic dysfunction, spongiform degeneration, astrocytic gliosis, and possibly PrP amyloid plaques (3,16,19,156,158,160,161,176,178,193,195–199).

structural and tinctorial features of amyloid (103). Furthermore, dispersion of prion rods into DLPCs results in a 10- to 100-fold increase in scrapie titer, and no rods could be identified in these fractions by electron microscopy (87).

XIII. DISTINCT PRION ISOLATES OR "STRAINS"

While a wealth of data argues that PrP^{Sc} alone can transmit scrapie prion infectivity, the molecular basis of prion diversity remains enigmatic (188). Dickinson, Kimberlin, and colleagues have provided convincing evidence for the existence of distinct prion isolates or "strains" with different properties (79,80). Prion "strains" have been characterized by their incubation times and neuropathologic lesion profiles in mice and hamsters.

Studies with Tg(SHaPrP) mice have shown that the PrP gene and PrP^{Sc} in the inoculum govern scrapie incubation times, neuropathology, and prion synthesis (Fig. 3) (174). These results suggest that transgenetic studies may provide a unique opportunity to dissect the molecular basis of scrapie "strains." Results with two isolates or "strains" of SHa prions inoculated into Tg(SHaPrP) mice argue that the expression and metabolism of PrP may profoundly influence the isolate phenotype. Sc237 (similar to 263 K) isolate of SHa prions produce 77 ± 1 day (n = 48) incubation times in Syrian hamsters while the 139H isolate yields 168 ± 7 day (n = 54) incubation times (80). $SHaPrP^C$ expression in Tg(SHaPrP)7 mice, about four- to eight-fold higher than in Syrian hamsters, gives incubation times of 48 ± 1 days (n = 26) with Sc237 and 40 ± 3 days (n = 11) with 139H (189). One interpretation of these observations is that Sc237 prions have a higher affinity for PrP^{Sc} than 139H prions. Increased expression of PrP^C substrate in Tg(SHaPrP) mice might saturate the PrP^{Sc} conversion process and result in diminution of the incubation times for both prion isolates and in obliteration of the differences between the two. In Chinese and Armenian hamsters whose PrP gene sequences differ from that of the Syrian at codons 7 and 8, respectively (118), 139H produces either shorter or similar incubation times to those observed with Sc237. In this case, the amino acid sequence of PrP may modulate the affinities of PrP^{Sc} in the two isolates for PrP^C molecules; indeed, the formation of PrP^C/PrP^{Sc} heterodimers may be the rate-limiting step in prion biosynthesis and thus may determine scrapie incubation times.

Defining the molecular basis of "prion strains" is one of the most challenging perplexing problems in biology. A variety of hypotheses have been offered to explain distinct isolates of prions. Despite the lack of physical, chemical, and biological evidence for a scrapie-specific nucleic acid, the remote scenario that such a molecule exists remains a formal possibility to explain the distinct prion isolates. A second hypothesis suggests that PrP^{Sc} alone can cause scrapie with the properties of Sc237 but that the 139H isolate contains an accessory RNA molecule which modulates the properties of this prion (188). Such accessory RNAs are of cellular origin and thus would not be detected by subtractive hybridization and differential cloning studies. Synthesis of such RNA molecules would be triggered by PrP^{Sc} or the hypothetical accessory RNA in the prion inoculum. A third hypothesis considers the possibility that scrapie ''strains'' may have their origin in a non-PrP molecule which purifies with PrP^{Sc} but it is not a nucleic acid and as yet it is undetected. A fourth possibility is that different chemical or conformational modifications of PrP^{Sc} are responsible for the particular biological properties exhibited by scrapie prion ''strains'' (1).

XIV. SIGNIFICANCE OF PRION RESEARCH

Defining the complete molecular structure of the infectious prion particle and learning how prions replicate lie at the center of both basic and applied future advances in prion biology. Whether prions are composed entirely of PrP^{Sc} molecules or contain a second component needs to be resolved. Learning the molecular events which feature in prion replication should help decipher the structural basis for scrapie isolates or ''strains'' which give different incubation times in the same host. Elucidating the function of PrP^{C} might extend our understanding of the pathogenesis of prion diseases and provide clues to other macromolecules which may participate in a variety of human and animal diseases of unknown etiology. It seems likely that lessons learned from prion diseases may give insights into the etiologies and pathogenic mechanisms of such common CNS degenerative disorders afflicting older people as Alzheimer's disease, amyotrophic lateral sclerosis, and Parkinson's disease.

The advances in our knowledge of prions derived from mice that express foreign and mutant PrP transgenes demand the production and analysis of many new lines of Tg(PrP) mice; however, such research is necessarily limited by the economics of prolonged bioassays. Transgenetics offers new approaches to dissecting the mechanism of prion replication and to deciphering the enigma posed by ''strains'' of prions. PrP gene targeting has produced mice with both PrP alleles ablated. These mice with ablated PrP genes will greatly enhance transgenetic studies since they provide a method for removal of the host MoPrP genes while leaving the PrP transgene of interest.

106   STANLEY B. PRUSINER

Although the results of many studies indicate that prions are a new class of pathogens distinct from both viroids and viruses, it is unknown whether different types of prions exist. Are there prions that contain modified proteins other than PrP^{Sc}? Assessing how widespread prions are in nature and defining their subclasses are subjects for future investigation. Elucidation of the mechanism by which, in prion diseases, brain cells, after a long delay, cease to function and die may offer approaches to understanding how neurons develop, mature, transmit signals for decades, and eventually grow senescent.

Acknowledgments

Portions of this manuscript were adapted from an article entitled "Molecular biology of prion diseases" published in *Science* 252:1515–1522, 1991. This work was supported by grants from the National Institutes of Health and the American Health Assistance Foundation, as well as by gifts from the Sherman Fairchild Foundation and National Medical Enterprises.

References

1. Prusiner, S.B. (1991). *Science* **252,** 1515–1522.
2. Prusiner, S.B. (1982). *Science* 216, 136–144.
3. Hsiao, K., Baker, H.F., Crow, T.J., Poulter, M., Owen, F. Terwilliger, J.D., Westaway, D., Ott, J., and Prusiner, S.B. (1989). *Nature* **338,** 342–345.
4. Wells, G.A.H., Scott, A.C., Johnson, C.T., Gunning, R.F., Hancock, R.D., Jeffrey, M., Dawson, M. and Bradley, R. (1987). A novel progressive spongiform encephalopathy in cattle. *Vet. Rec.* **121,** 419–420.
5. Wilesmith, J.W., Wells, G.A.H., Cranwell, M.P., and Ryan, J.B.M. (1988). *Vet. Rec.* **123,** 638–644.
6. Hope, J., Reekie, L.J.D., Hunter, N., Multhaup, G., Beyreuther, K., White, H., Scott, A.C., Stack, M.J., Dawson, M., and Wells, G.A.H. (1988). *Nature* **336,** 390–392.
7. Wilesmith, J., and Wells, G.A.H. (1991). *Curr. Top. Microbiol. Immunol.* **172,** 21–38.
8. Kirkwood, J.K., Wells, G.A.H., Wilesmith, J.W., Cunningham, A.A., and Jackson, S.I. (1990). *Vet. Rec.* **127,** 418–420.
9. Pain, S. (1990). *New Scientist,* 32–34.
10. Bradley, R. (1990). *J. Pathol.* **160,** 283–285.
11. Dealler, S.F., and Lacey, R.W. (1990). *Food Microbiol.* **7,** 253–279.
12. Winter, M.H., Aldridge, B.M., Scott, P.R., and Clarke, M. (1989). *Br. Vet. J.* **145,** 191–194.
13. Scott, A.C., Wells, G.A.H., Stack, M.J., White , H., and Dawson, M. (1990) *Vet. Microbiol.* **23,** 295–304.

14. Parry, H.B. (1983). *In* "Scrapie Disease in Sheep" (Oppenheimer D.R., ed.), p. 192, Academic Press, New York.
15. Parry, H.B. (1962). *Heredity* **17,** 75–105.
16. Gajdusek, D.C. (1977). *Science* **197,** 943–960.
17. Gajdusek, D.C., Gibbs, C.J., Jr., and Alpers, M. (1966). *Nature* **209,** 794–796.
18. Gibbs, C.J., Jr., Gajdusek, D.C., Asher, D.M., Alpers, M.P., Beck, E., Daniel, P.M., and Matthews, W.B. (1968). *Science* **161,** 388–389.
19. Masters, C.L., Gajdusek, D.C., and Gibbs, C.J., Jr. (1981). *Brain* **104,** 535–558.
20. Masters, C.L., Gajdusek, D.C., Gibbs, C.J., Jr. (1981). *Brain* **104,** 559–588.
21. Alpers, M. (1987) *In* "Prions—Novel Infectious Pathogens Causing Scrapie and Creutzfeldt-Jakob Disease" (S.B. Prusiner, and M.P. McKinley, eds.), pp. 451–465, Academic Press, Orlando.
22. Gordon, W.S. (1996). *Vet. Res.* **58,** 516–520.
23. Chandler, R.L. *Lancet* **1,** 1378–1379.
24. Alper, T., Haig, D.A., and Clarke, M.C. (1966). *Biochem. Biophys. Res. Commun.* **22,** 278–284.
25. Alper, T, Cramp, W.A., Haig, D.A., and Clarke, M.C. (1967). *Nature* **214,** 764–766.
26. Hunter, G.D. (1972) *J. Infect. Dis.* **125,** 427–440.
27. Latarjet, R., Muel, B., Haig, D.A., Clarke, M.C., and Alper, T. (1970). *Nature* **227,** 1341–1343.
28. Alper, T., Haig, D.A., and Clarke, M.C. (1978) *J. Gen. Virol.* **41,** 503–516.
29. Gibbons, R.A., and Hunter, G.D. (1967). *Nature* **215,** 1041–1043.
30. Griffith, J.S. (1967). *Nature* **215,** 1043–1044.
31. Millson, G., Hunter, G.D., and Kimberlin, R.H. (1971). *J. Comp. Pathol.* **81,** 255–265.
32. Pattison, I.H., Jones, K.M. (1967) *Vet. Rec.* **80,** 1–8.
33. Prusiner, S.B., Groth, D.F., Cochran, S.P., Masiarz, F.R., McKinley, M.P., and Martinez, H.M. (1980). *Biochemistry* **19,** 4883–4891.
34. Prusiner, S.B., Cochran, S.P., Groth, D.F., Downey, D.E., Bowman, K.A., and Martinez, H.M. (1982). *Ann. Neurol.* **11,** 353–358.
35. Prusiner, S.B., McKinley, M.P., Groth, D.F., Bowman, K.A., Mock, N.I., Cochran, S.P., and Masiarz, F.R. (1981). *Proc. Natl. Acad. Sci. U.S.A.* **78,** 6675–6679.
36. Hunter, G.D., and Millson, G.C. (1967). *J. Comp. Pathol.* **77,** 301–307.
37. Hunter, G.D., Gibbons, R.A., Kimberlin, R.H., and Millson, G.C. (1969). *J. Comp. Pathol.* **79,** 101–108.
38. Cho, H.J. (1980). *Intervirology* **14,** 213–216.
39. Diener, T.O., McKinley, M.P. and Prusiner, S.B. (1982). *Proc. Natl. Acad. Sci. U.S.A.* **79,** 5220–5224.
40. McKinley, M.P., Masiarz, F.R., Isaacs, S.T., Hearst, J.E., and Prusiner, S.B. (1983). *Photochem. Photobiol.* **37,** 539–545.
41. Bellinger-Kawahara, C., Cleaver, J.E., Diener, T.O., and Prusiner, S.B. (1987). *J. Virol.* **61,** 159–166.
42. Bellinger-Kawahara, C., Diener, T.O., McKinley, M.P., Groth, D.F., Smith, D.R., and Prusiner, S.B. (1987). *Virology* **160,** 271–274.
43. Gabizon, R., McKinley, M.P., Groth, D.F., Kenaga, L., and Prusiner, S.B. (1988). *J. Biol. Chem.* **263,** 4950–4955.

44. Bolton, D.C., McKinley, M.P., and Prusiner, S.B. (1982). *Science* **218**, 1309–1311.
45. Prusiner, S.B., Bolton, D.C., Groth, D.F., Bowman, K.A., Cochran, S.P., and McKinley, M.P. (1982). *Biochemistry* **21**, 6942–6950.
46. McKinley, M.P., Bolton, D.C., and Prusiner, S.B. (1983). *Cell* **35**, 57–62.
47. Prusiner, S.B., Groth, D.F., Bolton, D.C., Kent, S.B., and Hood, L.E. (1984). *Cell* **38**, 127–134.
48. Oesch, B., Westaway, D., Wälchli, M., McKinley, M.P., Kent, S.B.H., Aebersold, R., Barry, R.A., Tempst, P., Teplow, D.B., Hood, L.E., Prusiner, S.B., and Weissmann, C. (1985). *Cell* **40**, 735–746.
49. Basler, K., Oesch, B., Scott, M., Westaway, D., Wälchli, M., Groth, D.F., McKinley, M.P., Prusiner, S.B., and Weissmann, C. (1986). *Cell* **46**, 417–428.
50. Chesebro, B., Race, R., Wehrly, K., Nishio, J., Bloom, M., Lechner, D., Bergstrom, S., Robbins, K., Mayer, L., Keith, J.M., Garon, C., and Haase, A. (1985). *Nature* **315**, 331–333.
51. Locht, C., Chesebro, B., Race, R. and Keith, J.M. (1986). *Proc. Natl. Acad. Sci. U.S.A.* **83**, 6372–6376.
52. Kretzschmar, H.A., Stowring, L.E., Westaway, D., Stubblebine, W.H., Prusiner, S.B., and DeArmond, S.J. (1986). *DNA* **5**, 315–324.
53. Hope, J., Morton, L.J.D., Farquhar, C.F., Multhaup, G., Beyreuther, K., and Kimberlin, R.H. (1986). *EMBO J.* **5**, 2591–2597.
54. Turk, E., Teplow, D.B., Hood, L.E., and Prusiner, S.B. (1988). *Eur. J. Biochem.* **176**, 21–30.
55. Safar, J., Wang, W., Padgett, M.P., Ceroni, M., Piccardo, P., Zopf, D., Gajdusek, D.C., and Gibbs, C.J., Jr. *Proc. Natl. Acad. Sci. U.S.A. 87*, 6373–6377.
56. Stahl, N., Borchelt, D.R., Hsiao, K., and Prusiner, S.B. (1987). *Cell* **51**, 229–240.
57. Stahl, N., Baldwin, M.A., Burlingame, A.L., and Prusiner, S.B. (1990) *Biochemistry* **29**, 8879–8884.
58. Baldwin, M.A., Stahl, N., Reinders, L.G., Gibson, B.W., Prusiner, S.B., and Burlingame, A.L. (1990). *Anal. Biochem.* **191**, 174–182.
59. Safar, J., Ceroni, M., Piccardo, P., Liberski, P.P., Miyazaki, M., Gajdusek, D.C., and Gibbs, C.J., Jr. (1990). *Neurology* **40**, 503–508.
60. Bolton, D.C., Meyer, R.K., and Prusiner, S.B. (1985). *J. Virol.* **53**, 596–606.
61. Manuelidis, L., Valley, S., and Manuelidis, E.E. (1985). *Proc. Natl. Acad. Sci. U.S.A.* **82**, 4263–4267.
62. Haraguchi, T., Fisher, S., Olofsson, S., Endo, T., Groth, D., Tarantino, A., Borchelt, D.R., Teplow, D., Hood, L., Burlingame, A., Lycke, E., Kobata, A., and Prusiner, S.B. (1989). *Arch. Biochem. Biophys. 274*, 1–13.
63. Endo, T., Groth, D., Prusiner, S.B., and Kobata, A. *Biochemistry* **28**, 8380–8388.
64. Rogers, M., Taraboulos, A., Scott, M., Groth, D., and Prusiner, S.B. (1990). *Glycobiology* **1**, 101–109.
65. Stahl, N., Baldwin, M.A., Teplow, D., Hood, L.E., Beavis, R., Chait, B., Gibson, B., Burlingame, A.L., and Prusiner, S.B. (1992). *In* "Prion Diseases in Humans and Animals" (S.B. Prusiner, J. Collinge, J. Powell, B. Anderton, eds.),pp. 361–379, Ellis Horwood, London.
66. Braig, H., and Diringer, H. (1985). *EMBO J.* **4**, 2309–2312.
67. Sklaviadis, T.K., Manuelidis, L., and Manuelidis, E.E. (1989). *J. Virol.* **63**, 1212–1222.

68. Sklaviadis, T., Akowitz, A., Manuelidis, E.E., and Manuelidis, L. (1990). *Arch. Virol.* **112,** 215–229.
69. Aiken, J.M., Williamson, J.L., and Marsh, R.F. (1989). *J. Virol.* **63,** 1686–1694.
70. Aiken, J.M., Williamson, J.L., Borchardt, L.M., and Marsh, R.F. (1990). *J. Virol.* **64,** 3265–3268.
71. Akowitz, A., Sklaviadis, T., Manuelidis, E.E., and Manuelidis, L. (1990). *Microb. Pathog.* **9,** 33–45.
72. Murdoch, G.H., Sklaviadis, T., Manuelidis, E.E., and Manuelidis, L. (1990). *J. Virol.* **64,** 1477–1486.
73. Manuelidis, L., and Manuelidis, E.E. (1989). *Microb. Pathog.* **7,** 157–164.
74. Czub, M., Braig, H.R., and Diringer, H. (1986). *J. Gen. Virol.* **67,** 2005–2009.
75. Czub, M., Braig, H.R., and Diringer, H. (1988). *J. Gen. Virol.* **69,** 1753–1756.
76. Jendroska, K., Heinzel, F.P., Torchia, M., Stowring, L., Kretzschmar, H.A., Kon, A., Stern, A., Prusiner, S.B., and DeArmond, S.J. (1991). *Neurology* **41,** 1482–1490.
77. Pocchiari, M. (1990). International Symposium on Virological Aspects of the Safety of Biological Products, London, England, Nov. 8–9 (Abstr.) 1990; p. 13, International Association of Biological Standardization.
78. Dickinson, A.G., and Fraser, H. (1979). *In* ''Slow Transmissible Diseases of the Nervous System, Vol. 1.'' (S.B. Prusiner, and W.J. Hadlow, eds.), pp. 367–386, *Academic Press,* New York.
79. Bruce, M.E., and Dickinson, A.G. (1987). *J. Gen. Virol.* **68,** 79–89.
80. Kimberlin, R.H., Cole, S., and Walker, C.A. (1987). *J. Gen. Virol.* **68,** 1875–1881.
81. Dickinson, A.G., and Outram, G.W. (1988). *In* ''Novel Infectious Agents and the Central Nervous System. Ciba Foundation Symposium 135'' (G. Bock and J. Marsh, eds.), pp. 63–83, John Wiley & Sons, Chichester, UK.
82. Weitgrefe, S., Zupancic, M., Haase, A., Chesebro, B., Race, R., Frey, W., II, Rustan, T., Friedman, R.L. (1985). *Science* **230,** 1177–1181.
83. Diedrich, J., Weitgrefe, S., Zupancic, M., Staskus, K., Retzel, E., Haase, A.T., and Race, R. (1987). *Microb. Pathog.* **2,** 435–442.
84. Duguid, J.R., Rohwer, R.G., and Seed, B. (1988). *Proc. Natl. Acad. Sci. U.S.A.* **85,** 5738–5742.
85. Oesch, B., Groth, D.F., Prusiner, S.B., and Weissmann, C. (1988). *In* ''Novel Infectious Agents and the Central Nervous System, Ciba Foundation Symposium 135,'' (G. Bock, and J. Marsh, eds.), pp. 209–223, John Wiley & Sons, Chichester, U.K.
86. Meyer, N., Rosenbaum, V., Schmidt, B., Gilles, K., Mirenda, C., Groth, D., Prusiner, S.B., Riesner, D. (1991). *J. Gen. Virol.* **72,** 37–49.
87. Gabizon, R., McKinley, M.P., and Prusiner, S.B. (1987). *Proc. Natl. Acad. Sci. U.S.A.* **84,** 4017–4021.
88. Prusiner, S.B., McKinley, M.P., Bowman, K.A., Bolton, D.C. Bendheim, P.E., Groth, D.F., Glenner, and G.G. (1983). *Cell* **35,** 349–358.
89. Bendheim, P.E., Barry, R.A., DeArmond, S.J., Stites, D.P., and Prusiner, S.B. (1984). *Nature* **310,** 418–421.
90. DeArmond, S.J., McKinley, M.P., Barry, R.A., Braunfeld, M.B., McColloch, J.R., and Prusiner, S.B. (1985). *Cell* **41,** 221–235.
91. Kitamoto, T., Tateishi, J., Tashima, I., Takeshita, I., Barry, R.A., DeArmond, S.J., and Prusiner, S.B. (1986). *Ann. Neurol.* **20,** 204–208.

92. Roberts, G.W., Lofthouse, R., Allsop, D., Landon, M., Kidd, M., Prusiner, S.B., and Crow, T.J. (1988). *Neurology* **38**, 1534–1540.

93. Tagliavini, F., Prelli, F., Ghisto, J., Bugiani, O., Serban, D., Prusiner, S.B., Farlow, M.R., Ghetti, B., and Frangione, B. (1991). *EMBO J.* **10**, 513–519.

94. Merz, P.A., Rohwer, R.G., Kascsak, R., Wisniewski, H.M., Somerville, R.A., Gibbs, C.J., Jr., and Gajdusek, D.C. (1984). *Science* **225**, 437–440.

95. Merz, P.A., Somerville, R.A., Wisniewski, H.M., and Iqbal, K. (1981). *Acta Neuropathol. (Berl.)* **54**, 63–74.

96. Merz, P.A., Wisniewski, H.M., Somerville, R.A., Bobin, S.A., Masters, C.L., and Iqbal, K. (1983). *Acta Neuropathol. (Berl.)* **60**, 113–124.

97. Diringer, H., Gelderblom, H., Hilmert, H., Ozel, M., Edelbluth, C., and Kimberlin, R.H. (1983). *Nature* **306**, 476–478.

98. Merz, P.A., Kascsak, R.J., Rubenstein, R., Carp, R.I., and Wisniewski, H.M. (1987). *J. Virol.* **61**, 42–49.

99. Somerville, R.A., Ritchie, L.A., and Gibson, P.H. (1989). *J. Gen. Virol.* **70**, 25–35.

100. Diener, T.O. (1987). *Cell* **49**, 719–721.

101. Kimberlin, R.H. (1990). *Semin. Virol.* **1**, 153–162.

102. Aiken, J.M., and Marsh, R.F. (1990). *Microbiol. Rev.* **54**, 242–246.

103. McKinley, M.P., Meyer, R., Kenaga, L., Rahbar, F., Cotter, R., Serban, A., and Prusiner, S.B. (1991). *J. Virol.* **65**, 1440–1449.

104. Gabizon, R., McKinley, M.P., Groth, D.F., and Prusiner, S.B. (1988). *Proc. Natl. Acad. Sci. U.S.A.* **85**, 6617–6621.

105. Gabizon, R., and Prusiner, S.B. (1990). *Biochem. J.* **266**, 1–14.

106. Sparkes, R.S., Simon, M., Cohn, V.H., Fournier, R.E.K., Lem, J., Klisak, I., Heinzmann, C., Blatt, C., Lucero, M., Mohandas, T., DeArmond, S.J., Westaway, D., Prusiner, S.B., and Weiner, L.P. (1986). *Proc. Natl. Acad. Sci. U.S.A.* **83**, 7358–7362.

107. Liao, Y.-C., Lebo, R.V., Clawson, G.A., and Smuckler, E.A. (1986). *Science* **233**, 364–367.

108. Robakis, N.K., Devine-Gage, E.A., Kascsak, R.J., Brown, W.T., Krawczun, C., and Silverman, W.P. (1986). *Biochem. Biophys. Res. Commun.* **140**, 758–765.

109. Westaway, D., Goodman, P.A. Mirenda, C.A., McKinley, M.P., Carlson, G.A., and Prusiner, S.B. (1987). *Cell* **51**, 651–662.

110. Westaway, D., Mirenda, C.A., Foster, D., Zebarjadian, Y., Scott, M., Torchia, M., Yang, S.-L., Serban, H., DeArmond, S.J., Ebeling, C., Prusiner, S.B., and Carlson, G.A. (1991). *Neuron* **7**, 59–68.

111. Blum, B., Bakalara, N., and Simpson, L. (1990). *Cell* **60**, 189–198.

112. Kane, P.M., Yamashiro, C.T., Wolczyk, D.F., Neff, N., Goebl, M., and Stevens, T.H. (1990). *Science* **250**, 651–657.

113. McKnight, S., and Tjian, R. (1986). *Cell* **46**, 795–805.

114. Mobley, W.C., Neve, R.L., Prusiner, S.B., and McKinley, M.P. (1988). *Proc. Natl. Acad. Sci. U.S.A.* **85**, 9811–9815.

115. Kretzschmar, H.A., Prusiner, S.B., Stowring, L.E., and DeArmond, S.J. (1986). *Am. J. Pathol.* **122**, 1–5.

116. Goldman, W., Hunter, N., Foster, J.D. Salbaum, J.M., Beyreuther, K., and Hope, J. (1990). *Proc. Natl. Acad. Sci. U.S.A.* **87**, 2476–2480.

117. Goldman, W., Hunter, N., Manson, J., and Hope, J. (1990). VIIIth International Congress of Virology, Berlin, Aug. 26–31 (Abstr.) 284.
118. Lowenstein, D.H., Butler, D.A., Westaway, D., McKinley, M.P., DeArmond, S.J., and Prusiner, S.B. (1990). *Mol. Cell. Biol.* **10**, 1153–1163.
119. Harris, D.A., Falls, D.L., Walsh, W., and Fischbach, G.D. (1989). *Soc. Neurosci.* **15**, 70–77.
120. Falls, D.L., Harris, D.A., Johnson, F.A., Morgan, M.M., Corfas, G., and Fischbach, G.D. (1990). Cold Spring Harbor Sympos. Quant. Biol. **55**, 397–406.
121. Borchelt, D.R., Scott, M., Taraboulos, A., Stahl, N., and Prusiner, S.B. (1990). *J. Cell Biol.* **110**, 743–752.
122. Caughey, B., Race, R.E., Ernst, D., Buchmeier, M.J., and Chesebro, B. (1989). *J. Virol.* **63**, 175–181.
123. Caughey, B., and Raymond, G.J. (1991). *J. Biol. Chem.* **266**, 18217–18223.
124. Stahl, N., Borchelt, D.R., and Prusiner, S.B. (1990). *Biochemistry* **29**, 5405–5412.
125. Taraboulos, A., Serban, D., and Prusiner, S.B. (1990). *J. Cell Biol.* **110**, 2117–2132.
126. McKinley, M.P., Taraboulos, A., Kenaga, L., Serban, D., Stieber, A., DeArmond, S., Prusiner, S.B., and Gonatas, N. (1991). *Lab. Invest.* **65**, 622–630.
127. Butler, D.A., Scott, M.R.D., Bockman, J.M., Borchelt, D.R., Taraboulos, A., Hsiao, K.K., Kinsbury, D.T., and Prusiner, S.B. (1988). *J. Virol.* **62**, 1558–1564.
128. Taraboulos, A., Rogers, M., Borchelt, D.R., McKinley, M.P., Scott, M., Serban, D., and Prusiner, S.B. (1990). *Proc. Natl. Acad. Sci. U.S.A.* **87**, 8262–8266.
129. Yost, C.S., Lopez, C.D., Prusiner, S.B., Meyers, R.M., and Lingappa, V.R. (1990). *Nature* **343**, 669–672.
130. Lopez, C.D., Yost, C.S., Prusiner, S.B., Myers, R.M., and Lingappa, V.R. (1990). *Science* **248**, 226–229.
131. Hay, B., Prusiner, S.B., and Lingappa, V.R. (1987). *Biochemistry* **26**, 8110–8115.
132. Hay, B., Barry, R.A., Lieberburg, I., Prusiner, S.B., and Lingappa, V.R. (1987). *Mol. Cell. Biol.* **7**, 914–920.
133. Bazan, J.F., Fletterick, R.J., McKinley, M.P., and Prusiner, S.B. (1987). *Protein Eng.* **1**, 125–135.
134. Doms, R.W., Russ, G., and Yewdell, J.W. (1989). *J. Cell. Biol.* **109**, 61–72.
135. Lippincott-Schwartz, J., Yuan, L.C., Bonifacino, J.S., and Klausner, R.D. (1989). *Cell* **56**, 801–813.
136. Taraboulos, A., Raeber, A., Borchelt, D., McKinley, M.P., and Prusiner, S.B. (1991). *FASEB J.* **5**, A1177.
137. Prusiner, S.B., Hadlow, W.J., Garfin, D.E., Cochran, S.P., Baringer, J.R., Race, R.E., and Eklund, C.M. (1978). *Biochemistry* **17**, 4993–4997.
138. Carlson, G.A., Kingsbury, D.T., Goodman, P.A., Coleman, S., Marshall, S.T., DeArmond, S.J., Westaway, D., and Prusiner, S.B. (1986). *Cell* **46**, 503–511.
139. Hunter, N., Hope, J., McConnell, I., and Dickinson, A.G. (1987). *J. Gen. Virol.* **68**, 2711–2716.
140. Race, R.E., Graham, K., Ernst, D., Caughey, B., and Chesebro, B. (1990). *J. Gen. Virol.* **71**, 493–497.
141. Carlson, G.A., Goodman, P.A., Lovett, M., Taylor, B.A., Marshall, S.T., Peterson-Torchia, M., Westaway, D., and Prusiner, S.B. (1988). *Mol. Cell. Biol.* **8**, 5528–5540.
142. Dickinson, A.G., Meikle, V.M.H., and Fraser, H. (1968). *J. Comp. Pathol.* **78**, 293–299.

143. Dickinson, A.G., Young, G.B., Stamp, J.T., and Renwick, C.C. (1965). *Heredity* **20,** 485–503.
144. Hunter, N., Foster, J.D., Dickinson, A.G., and Hope, J. (1989). *Vet. Rec.* **124,** 364–366.
145. Owen, F., Poulter, M., Lofthouse, R., Collinge, J., Crow, T.J., Risby, D., Baker, H.F., Ridley, R.M., Hsiao, K., and Prusiner, S.B. (1989). *Lancet* **1,** 51–52.
146. Owen, F., Poulter, M., Shah, T., Collinge, J., Lofthouse, R., Baker, H., Ridley, R., McVey, J., and Crow, T. (1990). *Mol. Brain Res.* **7,** 273–276.
147. Collinge, J., Harding, A.E., Owen, F., Poulter, M., Lofthouse, R., Boughey, A.M., Shah, T., and Crow, T.J. (1989). *Lancet* **2,** 15–17.
148. Collinge, J., Owen, F., Poulter, H., Leach, M., Crow, T., Rosser, M., Hardy, J., Mullan, H., Janota, I., and Lantos, P. (1990). *Lancet* **336,** 7–9.
149. Crow, T.J., Collinge, J., Ridley, R.M., Baker, H.F., Lofthouse, R., Owen, F., and Harding, A.E. (1990). Seminar on Molecular Approaches to Research in Spongiform Encephalopathies in Man, Medical Research Council, London (Abstr.) 1990.
150. Owen, F., Poulter, M., Collinge, J., Leach, M., Shah, T., Lofthouse, R., Chen, Y.F., Crow, T.J., Harding, A.E., and Hardy, J. (1991). *Exp. Neurol.* **112,** 240–242.
151. Laplanche, J.-L., Chatelain, J., Launay, J.-M., Gazengel, C., and Vidaud, M. (1990). *Nucleic Acids Res.* **18,** 6745.
152. Goldfarb, L.G., Brown, P., McCombie, W.R., Goldgaber, D., Swergold, G.D., Wills, P.R., Cervenakova, L., Baron, H., Gibbs, C.J.J., and Gajdusek, D.C. (1991). *Proc. Natl. Acad. Sci. U.S.A.* **88,** 10926–10930.
153. Kahana, E., Milton, A., Braham, J., and Sofer, D. *Science* **183,** 90–91.
154. Alter, M., and Kahana, E. (1976). *Science* **192,** 428.
155. Herzberg, L., Herzberg, B.N., Gibbs, C.J., Jr., Sullivan, W., Amyx, H., and Gajdusek, D.C. (1974). *Science* **186,** 848.
156. Goldfarb, L.G., Mitrova, E., Brown, P., Toh, B.H., and Gajdusek, D.C. (1990). *Lancet* **336,** 514–515.
157. Gabizon, R., Meiner, Z., Cass, C., Kahana, E., Kahana, I., Avrahami, D., Abramsky, O., Scarlato, G., Prusiner, S.B., and Hsiao, K.K. (1991). *Neurology* **41,** 160 (abstr.).
158. Hsiao, K., Meiner, Z., Kahana, E., Cass, C., Kahana, I., Avrahami, D., Scarlato, G., Abramsky, O., Prusiner, S.B., and Gabizon, R. (1991). *N. Engl. J. Med.* **324,** 1091–1097.
159. Wexler, N.S., Young, A.B., Tanzi, R.E., Travers, H., Starosta-Rubinstein, S., Penney, J.B., Snodgrass, S.R., Shoulson, I., Gomez, F., Ramos Arroyo, M.A., Penchaszadeh, G.K., Moreno, H., Gibbons, K., Faryniarz, A., Hobbs, W., Anderson, M.A., Bonilla, E., Conneally, P.M., and Gusella, J.F. (1987). *Nature* **326,** 194–197.
160. Doh-ura, K., Tateishi, J., Sasaki, H., Kitamoto, T., and Sakaki, Y. (1989). *Biochem. Biophys. Res. Commun.* **163,** 974–979.
161. Hsiao, K.K., Cass, C., Schellenberg, G.D., Bird, T., Devine-Gage, E., and Prusiner, S.B. (1991). *Neurology* **41,** 681–684.
162. Goldfarb, L.G., Haltia, M., Brown, P., Nieto, A., Kovanen, J., McCombie, W.R., Trapp, S., and Gajdusek, D.C. (1991). *Lancet* **337,** 425.
163. Hsiao, K., Dloughy, S., Ghetti, B., Farlow, M., Cass, C., Da Costa, M., Conneally, M., Hodes, M.E., and Prusiner, S.B. (1992). *Nature Genetics* **1,** 68–71.
164. Farlow, M.R., Yee, R.D., Dlouhy, S.R., Conneally, P.M., Azzarelli, B., and Ghetti, B. (1989). *Neurology* **39,** 1446–1452.

165. Ghetti, B., Tagliavini, F., Masters, C.L., Beyreuther, K., Giaccone, G., Verga, L., Farlo, M.R., Conneally, P.M., Dlouhy, S.R., Azzarelli, B., and Bugiani, O. (1989). *Neurology* **39**, 1453–1461.

166. Nochlin, D., Sumi, S.M., Bird, T.D., Snow, A.D., Leventhal, C.M., Beyreuther, K., and Masters, C.L. (1989). *Neurology* **39**, 910–918.

167. Giaccone, G., Tagliavini, F., Verga, L., Frangione, B., Farlow, M.R., Bugiani, O., and Ghetti, B. (1990). *Brain Res.* **530**, 325–329.

168. Owen, F., Poulter, M., Collinge, J., and Crow, T.J. (1990). *Am. J. Hum. Genet.* **46**, 1215–1216.

169. Fradkin, J.E., Schonberger, L.B., Mills, J.L., Gunn, W.J., Piper, J.M., Wysowski, D.K., Thomson, R., Durako, S., and Brown, P. (1991). *JAMA* **265**, 880–884.

170. Buchanan, C.R., Preece, M.A., and Milner, R.D.G. (1991). *Br. Med. J.* **302**, 824–828.

171. Collinge, J., Palmer, M.S., and Dryden, A.J. (1991). *Lancet* **337**, 1441–1442.

172. Palmer, M.S., Dryden, A.J., Hughes, J.T., and Collinge, J. (1991). *Nature* **352**, 340–342.

173. Hardy, J. (1991). Prion dimers—a deadly duo. Trends in Neurosciences 1991; 14:423–424.

174. Prusiner, S.B., Scott, M., Foster, D., Pan, K.-M., Groth, D., Mirenda, C., Torchia, M., Yang, S.-L., Serban, D., Carlson, G.A., Hoppe, P.C., Westaway, D., and DeArmond, S.J. (1990). *Cell* **63**, 673–686.

175. Hsiao, K.K., Scott, M., Foster, D., Groth, D.F., DeArmond, S.J., and Prusiner, S.B. (1990). *Science* **250**, 1587–1590.

176. Hsiao, K.K., Groth, D., Scott, M., Yang, S.-L., Serban, A., Rapp, D., Foster, D., Torchia, M., DeArmond, S.J., Prusiner, S.B. (1992). *In* "Prion Diseases of Humans and Animals" (S.B. Prusiner, J. Collinge, J. Powell, B. Anderton, eds.), pp 120–128, Ellis Horwood, London.

177. Neary, K., Caughey, B., Ernst, D., Race, R.E., and Chesebro, B. (1991). *J. Virol.* **65**, 1031–1034.

178. Hsiao, K., and Prusiner, S.B. (1990). *Neurology* **40**, 1820–1827.

179. Weissmann, C. (1991). *Nature* **349**, 569–571.

180. Pattison, I.H. (1965). *In* "Slow, Latent and Temperate Virus Infections, NINDB Monograph 2" (D.C. Gajdusek, C.J. Gibbs, Jr., and M.P. Alpers, eds.), pp. 249–257, U.S. Government Printing Office, Washington, DC.

181. Pattison, I.H. (1966). *Res. Vet. Sci.* **7**, 207–212.

182. Bockman, J.M., Prusiner, S.B., Tateishi, J., and Kingsbury, D.T. (1987). *Ann. Neurol.* **21**, 589–595.

183. Scott, M., Foster, D., Mirenda, C., Serban, D., Coufal, F., Wälchli, M., Torchia, M., Groth, D., Carlson, G., DeArmond, S.J., Westaway, D., and Prusiner, S.B. (1989). *Cell* **59**, 847–857.

184. Bellinger-Kawahara, C.G., Kempner, E., Groth, D.F., Gabizon, R., and Prusiner, S.B. (1988). *Virology* **164**, 537–541.

185. Gajdusek, D.C. (1988). *J. Neuroimmunol.* **20**, 95–110.

186. Gajdusek, D.C., (1990). *In* "Virology," 2nd ed. (B.N. Fields, D.M. Knipe, R.M. Chanock, M.S. Hirsch, J.L. Melnick, T.P. Monath, B. Roizman, eds.), pp. 2289–2324, Raven Press, New York.

187. Gajdusek, D.C., and Gibbs, C.J., Jr. (1990). *In* "Biomedical Advances in Aging" (A. Goldstein, ed.), pp. 3–24, Plenum Press, New York.

188. Weissmann, C. (1991). *Nature* **352**, 679–683.

189. Hecker, R., Stahl, N., Baldwin, M., Hall, S., McKinley, M.P., and Prusiner, S.B. (1990). VIIIth Intl. Congr. Virol., Berlin, Aug. 26–31, 1990; 284.

190. Bockman, J.M., Kingsbury, D.T., McKinley, M.P., Bendheim, P.E., and Prusiner, S.B. (1985). *N. Engl. J. Med.* **312**, 73–78.

191. Brown, P., Coker-Vann, M., Pomeroy, K., Franko, M., Asher, D.M., Gibbs, C.J., Jr., and Gajdusek, D.C. (1986). *N. Engl. J. Med.* **314**, 547–551.

192. Serban, D., Taraboulos, A., DeArmond, S.J., and Prusiner, S.B. (1990). *Neurology* **40**, 110–117.

193. Goldfarb, L., Korczyn, A., Brown, P., Chapman, J., and Gajdusek, D.C. (1990). *Lancet* **336**, 637–638.

194. Goldman, W., Hunter, N., Martin, T., Dawson, M., and Hope, J. (1991). *J. Gen. Virol.* **72**, 201–204.

195. Goldfarb, L.G., Brown, P., Goldgaber, D., Asher, D.M., Rubenstein, R., Brown, W.T., Piccardo, P., Kascsak, R.J., Boellaard, J.W., and Gajdusek, D.C. (1990). *Exp. Neurol.* **108**, 247–250.

196. Goldgaber, D., Goldfarb, L.G., Brown, P., Asher, D.M., Brown, W.T., Lin S., Teener, J.W., Feinstone, S.M., Rubenstein, R., Kascsak, R.J., Boellaard, J.W., and Gajdusek, D.C. (1989). *Exp. Neurol.,* **106**, 204–206.

197. Goldfarb, L., Brown, P., Goldgaber, D., Garruto, R., Yanaghiara, R., Asher, D., and Gajdusek, D.C. (1990). *Lancet* **336**, 174–175.

198. Hsiao, K.K., Doh-ura, K., Kitamoto, T., Tateishi, J., and Prusiner, S.B. (1989). *Ann. Neurol.* **26**, 137.

199. Tateishi, J., Kitamoto, T., Doh-ura, K., Sakaki, Y., Steinmetz, G., Tranchant, C., Warter, J.M., and Heldt, N. (1990). *Neurology* **40**, 1578–1581.

200. Borchelt, D.R., Taraboulos, A., Prusiner, S.B. (1992). *J. Biol. Chem.* **267**, 6188–6199.

201. Race, R.E., Fadness, L.H., Cheseboro, B. (1993). *J. Gen. Virol.* **68**, 1391–1399.

202. Prusiner, S.B., Fuzi, M., Scott, M., Serban, D., Serban, H., Taraboulos, A., Gabriel, J.-M., Wells, G., Wilesmith, J., Bradley, R., DeArmond, S.J., Kristensson, K. (1993). *J. Infect. Dis.,* **167**, 602–613.

203. Stahl, N., Baldwin, M.A., Teplow, D.B., Hood, L., Gibxon, B.W., Burlingame, A.L., Prusiner, S.B. (1993). *Biochemistry,* **32**, 1991–2002.

CATALYTIC ANTIBODIES

STEPHEN J. BENKOVIC

Department of Chemistry, The Pennsylvania State University,
University Park, Pennsylvania

I. Introduction

THE field of catalytic antibodies owes its origin to the hypothesis that enzymic catalysis derives from the complementarity of the enzyme's active site to the transition state or high-energy reaction intermediates generated in the course of a chemical transformation (Pauling, 1948). These interactions generated through hydrogen-bonding, and electrostatic and hydrophobic contacts, increase after the initial binding event of substrate and enzyme, until the transition state is reached; they then subsequently decrease permitting dissociation of the enzyme-product complex (Fig. 1). This concept, of course, applies to all catalytic phenomenon, but was extended specifically to the concept of a catalytic antibody by Pauling (1990) and Jencks (1969).

There were numerous attempts to induce catalytic antibodies—some published (Slobin, 1966; Raso and Stollar, 1975), others not (M. Cohn, personal communication). These failed, owing to the handicap of working with polyclonal mixtures (the active antibody catalyst would be an undetectable component) and an inadequate knowledge of reaction mechanisms. In the mid-1980s, the groups of Lerner (Tramontano et al., 1986) and Schultz (Pollack et al., 1986) demonstrated the induction of catalytic antibodies to, for want of a better description, transition state mimics of the tetrahedral species found in ester hydrolysis. We (Napper, 1987) demonstrated a similar induction of antibodies capable of catalyzing lactone formation, which showed, in addition, the feature of substrate enantioselectivity, a familiar characteristic of enzyme-catalyzed processes. There are now in excess of 50 different chemical reactions known to be catalyzed by antibodies (Lerner et al., 1991), and it is my intent, not to survey them exhaustively, but to group them into

115

The Harvey Lectures, Series 87, pages 115–128
© 1993 Wiley-Liss, Inc.

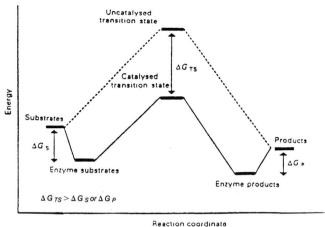

Fig. 1.   Reaction coordinate for uncatalyzed versus catalyzed reactions. The catalyst increases the rate of the chemical reaction by reducing the activation energy of
the reaction.

categories, to examine one in some mechanistic detail, and to then discuss
recent advances in the field. However, before commencing on that course, it
is pertinent to review briefly the chief structural features of antibodies.

II. Structural Characteristics of Antibodies

Antibodies are large proteins assembled by disulfide cross-links into a
four-chain structure (Fig. 2). The molecule consists of two identical heavy
chains of molecular weight approximately 50,000, and two identical light
chains of molecular weight 25,000 (Edelman et al., 1969). Sequence comparison of monoclonal immunoglobulin G (IgG) proteins indicates that the
carboxy-terminal half of the light chain and roughly three-fourths of the
carboxyl-end of the heavy chain show little sequence variation (Kabat et al.,
1983), whereas the amino-terminal regions of both chains (the first 100 amino
acids) show considerable sequence variability. Proteolytic cleavage on the
carboxy-terminal side of the interstrand disulfide linkage connecting the heavy
and light chains provides the means of generating molecules termed F_{ab}s containing a single antigen binding site (Fig. 2). A shortened version can be generated by recombinant methods, in which only hypervariable regions of the
heavy and light chains are joined by a short peptide sequence of 14 amino acids

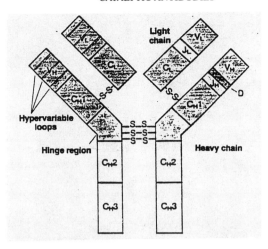

Fig. 2. The antibody binding site is encoded by the V_H, D, and J_H genes for the heavy chain and the V_L and J_L genes for the light chain, as shown on the right arm of this schematic representation of a typical antibody molecule. Within the variable regions are six hypervariable loops, three each in the light and heavy chains, as shown on the left arm of the diagram. The constant regions consist of C_L for the light chain and C_H1, C_H2, and C_H3 for the heavy chain as well as the hinge region. The heavy and light chains of the antibody molecule are held together by a series of disulfide bonds.

linking the carboxy-terminal end of the light chain with the amino-terminal end of the heavy chain. These are designated SCAs or single chain antibodies.

The X-ray crystallographic evidence reveals that all antibody structures are remarkably similar (Davies et al., 1990). There is a small repertoire of main-chain conformations—"canonical structures"—for at least five of the six variable regions of antibodies (Chothia et al, 1989). The conformation of these regions is determined by a few key residues. For small organic molecules the binding to antibody may occur by way of clefts whose volumes are in excess of 600 Å^3 (Getzoff et al., 1988), with dissociation constants for antibody-antigen union ranging from 10^{-4} to 10^{-14} M^{-1} (Watt et al., 1980). If binding were totally coupled to drive a chemical transformation, it would overcome a free energy change of up to 20 kcal M^{-1}. The binding of antigen does not result in a global conformational change in the antibody, but the union is accommodated by conformational adjustments in specific amino acids.

One may question whether the invariant nature of the antibody structure—a binding cavity composed of six variable loops on a β-sheet platform—will limit the scope of their reactivity. Insofar as the number and type of reactions catalyzed by antibodies continue to grow, one has some reassurance that this is not so.

III. Types of Antibody Catalysts

Free Energy Traps

Conceptually, the reactions most susceptible to antibody catalysis would be those which were originally viewed as "no-mechanism" reactions because their transition states manifest little polar or radical character. These transformations include electrocyclic reactions, sigmatropic rearrangements, and cycloadditions. Most of their transition states are cyclic in character and, therefore, amenable to acceleration by conformational or molecular restraint of the initial reactants.

One such process is the Diels-Alder reaction depicted in Figure 3. The transition state must more closely resemble the product than the conjugated diene and olefin starting materials (Sauer and Sustman, 1980). However, inducing antibodies to the product itself would produce binding sites subject to severe product inhibition. Two strategies have been created to circumvent this undesirable property: 1) catalyzing the formation of an initial, unstable bicyclic intermediate that as, in Figure 3 (top), extrudes SO_2 to yield the ultimate product (Hilvert et al., 1989), or 2) incorporating into the transition-state analog a molecular constraint that restricts the analog to a higher-energy conformational state than the product (Braisted and Schultz, 1990) (Fig. 3, bottom). Both strategies led to antibodies with multiple turnovers.

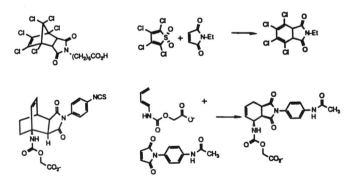

Fig. 3.   **Top and bottom:** Diels-Alder condensation. The inducing analogs are shown on the left. (Reproduced from Hilvert et al., 1989 [top] and Braisted and Schultz, 1990 [bottom], with permission of the publisher.)

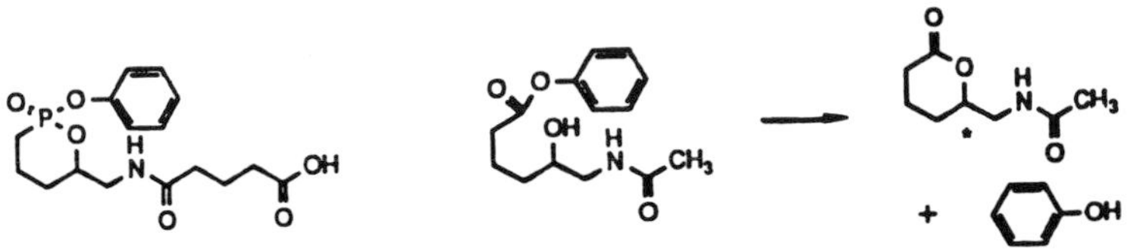

Fig. 4.   Lactonization (the same antibody catalyzes lactonization and amide bond formation. (Reproduced from Napper et al., 1987, with permission of the publisher.)

The intramolecular cyclization reaction featuring lactone formation (Fig. 4) presents an additional opportunity, namely for general acid-base catalysis, as well as a need to reduce the rotational entropy of the substrate in order to maximize the rate of lactonization. The ring closure in the absence of antibody is specific base catalyzed, consistent with nucleophilic attack by the alkoxide ion generated from the hydroxy ester. The pH versus rate profiles for k_{cat} and k_{cat}/K_M (the Michaelis-Menten parameters) for the antibody catalyzed process are first order in hydroxide ion (pH 7–10), and provide no evidence for the dissociation of a participating binding-site residue. This antibody catalyzed the enantioselective cyclization of one stereoisomer of the racemic hydroxy ester to give more than 94% enantiomeric excess.

From transition state theory, one can establish a reaction cycle (Fig. 5) to calculate the rate acceleration of the antibody-catalyzed reaction from the relative affinity of the antibody for the transition state versus the reactant (Benkovic et al., 1988). The values of K_A and K_L correspond to the Michaelis K_Ms for the substrates A and L (for an unimolecular process L = 1 M H_2O); the value of K_T is estimated from the K_i for the inducing hapten.

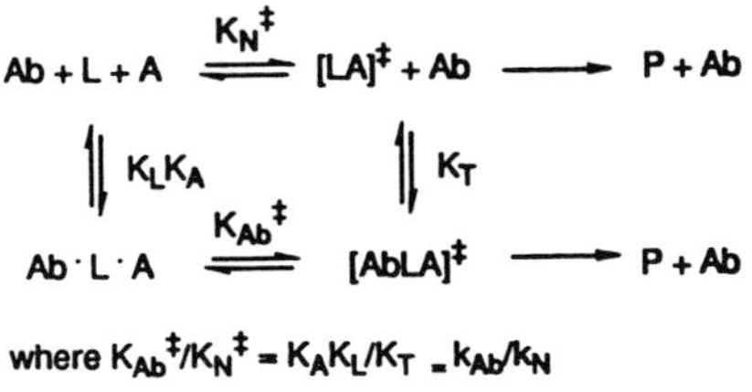

Fig. 5.   Reaction cycle to calculate the rate acceleration of the antibody-catalyzed reaction from the relative affinity of the antibody for the transition state versus the reactant. (Reproduced from Benkovic et al., 1988, with permission of the publisher.)

Insofar as the inducing hapten is a faithful transition-state mimic, the observed ratio of k_{ab}/k_N (the ratio of the rate coefficients for the antibody catalyzed reaction at substrate saturation to the spontaneous reaction of the substrates) is approximated; positive deviations suggest catalytic mechanisms more complex than those utilized by the antibodies described so far, whose behavior follows the relationship in Figure 5.

Active Catalysts

Since many chemical reactions are accelerated by general acid-base catalysis, the induction of desirable, appropriate functionalities has been attempted by a "bait and switch" process (Shokat et al., 1991; Janda et al., 1990; Lerner and Benkovic, 1988). Here, an additional residue (usually charged) is present in the immunogen but not in the substrate, so that the complementary residue induced in the antibody assumes an active role in catalysis. A pertinent example is depicted in Figure 6 in which the positively charged ammonium group is located so as to induce a carboxylate proximal to the α-proton that will be abstracted in the β-elimination process.

Chemical derivatization and genetic manipulation of binding site residues also hold promise for increasing the turnover numbers of catalytic antibodies, though to date the results are modest—at best a few hundred-fold (Pollack et al., 1988; Roberts et al., 1987). The problems of introducing and positioning these residues so that they function efficiently are akin to those encountered in the recombinant engineering of proteins, which has also not been spectacularly successful. However, it may prove easier to build in increased catalytic power in an induced site than to modify the active site of a highly efficient enzyme for another task. Thus, one may proceed along two parallel paths to obtain catalytic antibodies whose specificity and turnover numbers rival those of enzymes: 1) to screen thoroughly the enormous, diverse response of the immunological system for superior catalytic antibodies

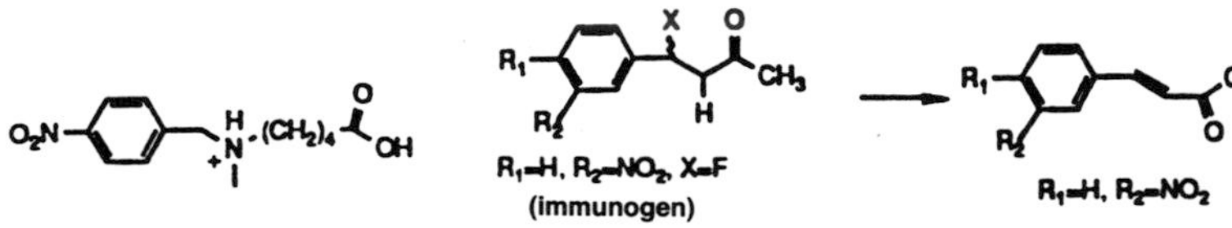

Fig. 6. β-Elimination. (Reproduced from Shokat and Schultz, 1991, with permission of the publisher.)

that possess, through chance assembly, a sophisticated active site; 2) to select a catalytic antibody with desired substrate specificity and modest catalytic activity and to improve it by random mutagenesis or combinatorial methods. Progress along either pathway is best served by mechanistic examination of existing catalytic antibodies, particularly those with high turnover numbers. I will describe one such example next.

IV. ANILIDE HYDROLYSIS BY 43C9

The antibody 43C9, obtained from screening some 50 monoclonal antibodies induced to a phosphonamide mimic of a tetrahedral intermediate, catalyzes the hydrolysis of a p-nitroanilide and a series of related esters (pNO$_2$, pCl, pCH$_3$, and pCH$_3$CO) (Fig. 7) (Janda et al., 1988; Benkovic et al., 1990). The steady-state Michaelis-Menten parameters, k_{cat}/K_M and k_{cat}, for both the p-nitroanilide and ester substrates are exhibited in Figure 8. At high pH the observed k_{cat} is more than 100 times that predicted by the above reaction cycle and approximately one million times faster than the spontaneous reaction rate. Consequently 43C9 is one of the most catalytically active antibodies known.

The data in Figure 8 have been fit to a mechanism that features a change in the rate-limiting step of an intermediate antibody-bound species due to changes in pH, rather than to a reaction mechanism involving the titration of a dissociable group at the antibody binding site. Support for the former sequence stems from several lines of evidence:

1. The plateau rate encountered in k_{cat} at pH >9.0 for hydrolysis of the p-nitroester can be assigned to the slow dissociation of the p-nitrophenol product.

2. The binding of p-nitrophenol and the p-nitroanilide substrate are pH-independent and do not reflect the pK_a in the k_{cat}/K_M profiles.

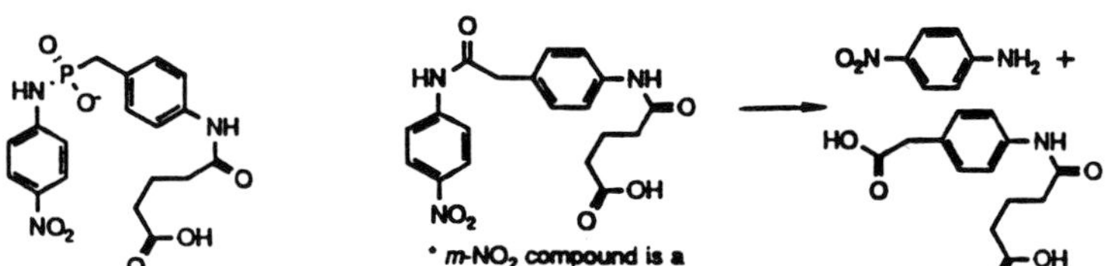

Fig. 7.   Amide hydrolysis (the same antibody catalyzes ester hydrolysis when the leaving alcohol is one of a series of p-substituted phenols. (Reproduced from Janda et al., 1988, with permission of the publisher.)

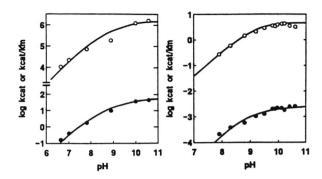

Fig. 8. pH-rate profiles for the hydrolysis of the *p*-nitrophenylester (left panel) and *p*-nitroanilide (right panel) by antibody 43C9 (see Fig. 7). (Reproduced from Benkovic et al., 1990, with permission of the publisher.)

3. There is a substantial deuterium solvent isotope effect (2–4) on the upward slopes of the pH-profiles but not in the plateau regions.

These observations collectively rule against a simple general base-catalyzed hydrolytic mechanism involving a single dissociable group, but are in accord with two different rate processes flanking an intermediate which defines the apparent pK_a. Efforts to detect the accumulation of an intermediate by rapid flow methods were unsuccessful. However, measurement of k_{cat} for the series of *p*-substituted esters gave a ρ value of $+2.3$ when correlated with the Hammett equation (Gibbs et al., 1992a); this is consistent with the ρ values of 2 to 3 observed for attack on acyl esters by nitrogen nucleophiles (Bruice and Mayahi, 1960), but not with the values of 1.0 to 1.2 and 0.5 to 0.7 found for nucleophilic attack by oxy anions or for general base catalysis (Kirsch et al., 1968).

The weight of this and other data (Benkovic et al., 1990) caused us to favor the reaction sequence shown in Figure 9, where S is either anilide or ester substrate, P_1 is the acid product, P_2 is the phenol or aniline, and I is the acylated antibody plus P_2. The pH versus rate profiles arise from a change in the rate-limiting step, from product release (ester) or acylation (anilide) at high pH to deacylation at low pH. The appearance of a large deuterium solvent isotope effect on the deacylation rates is not in accord with attack by hydroxide on an acyl intermediate, but suggests that a basic group on the antibody may be acting to catalyze deacylation. Its pK_a, like that of the nucleophile undergoing acylation, is outside the range of the pH versus rate profiles.

Conversion of 43C9 to a SCA form gave a catalyst with catalytic proper-

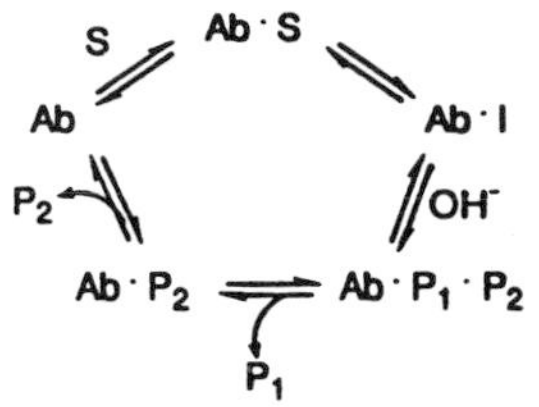

Fig. 9. Kinetic reaction sequence for ester and anilide hydrolysis catalyzed by 43C9. (Reproduced from Benkovic et al., 1990, with permission of the publisher.)

ties identical to those of the parent monoclonal antibody, and also provided the amino acid sequence of heavy and light chain regions (Gibbs et al., 1991). Molecular modeling (V.A. Roberts, unpublished results) of this sequence located two histidines in the combining site of 43C9, and furthermore suggested that one may be in a proper orientation for nucleophilic attack. In Table I are listed site-specific mutations to test the molecular model: H97Q (its imidazole is proximal to the carbonyl for acylation); H162N (its imidazole interacts with the aromatic ring of the substrate acid); and R102Q (its guanidinium group is within hydrogen-bonding distance of the oxygen of the carbonyl) (J. Stewart and N. Thomas, unpublished results). Conversion of histidine 97 to a glutamine abolishes catalytic activity but there is no change in hapten binding, while replacement of histidine 162 with an asparagine destroys catalytic activity but also weakens the hapten binding. We interpret these data in terms of H97 being an active site nucleophile and H162 being involved in maintenance of the antibody conformation necessary for productive binding. Substitution of arginine 102 by glutamine is likewise deleterious to catalytic activity and also weakens hapten binding. These findings are consistent with an important role for the guanidinium group of arginine 102 in electrostatic and hydrogen bond stabilization of the tetrahedral intermediates generated in the course of substrate hydrolysis, akin to the function of the oxyanion hole of the serine esterases.

Although we have recruited site-specific mutagenesis so far in a prospecting sense, once the validity of the model is confirmed, this procedure will be used, as noted earlier, to raise substrate turnover by 43C9 by further factor of 10^2 to 10^3 in order to rival the turnover numbers of the esterases.

One can question whether screening 43C9 from a small number of monoclonal antibodies selects a superior representative from the repertoire. The construction of bacteriophage libraries (Huse et al., 1989), using PCR

TABLE I. LIGAND BINDING TO MUTANT NPN 43C SINGLE-CHAIN ANTIBODIES

Protein	Catalytic?	Hapten		Acid		Phenol	
		K_D (nM)	ΔG (kcal/mole)	K_D (μM)	ΔG (kcal/mole)	K_D (μM)	ΔG (kcal/mole)
wt Mab	Yes	0.8 ± 0.2	−12.4 ± 0.3	28 ± 5	−6.2 ± 0.2	1.0 ± 0.2	−8.2 ± 0.2
wt	Yes	≤1	≤ −12.3	15 ± 1	−6.6 ± 0.1	0.6 ± 0.1	−8.5 ± 0.2
H97Q	No	≤1	≤ −12.3	N.D.	N.D.	N.D.	N.D.
H162N	No	16 ± 2	−10.6 ± 0.1	100 ± 10	−5.5 ± 0.1	N.D.	N.D.
R102Q	No	16 ± 4	−10.6 ± 0.3	31 ± 8	−6.1 ± 0.3	0.96 ± 0.12	−8.2 ± 0.1

N.D. = Not determined

Hapten

Acid

Phenol

(polymerase chain reaction) amplification of mRNA from immunized spleen cells or hybridomas to produce separate heavy-and light-chain libraries, followed by their random recombination into vectors that express $F_{ab}s$, offers a route to explore the rich diversity of the vast antibody repertoire. Typically combinatorial libraries of about one million clones expressing functional F_{ab} can be generated. After screening, selected clones are then overexpressed and assembled in *E. coli* (Better et al., 1988; Skerra and Pluckthun, 1988). We have recently generated an F_{ab} library of a million members from the spleen of a mouse immunized with the same phosphonamide hapten used to induce 43C9. Using vectors with improved expression, these libraries hold the potential of allowing direct screening, at the plaque level, for F_{ab} fragments with catalytic properties that exceed those of 43C9 (Posner et al., 1992).

V. Catalysis of a Multistep Reaction

Many reactions of chemical and biological interest are multistep, and by definition proceed through several transition states. A case in point is the conversion of a substrate containing an Asn-Gly sequence to the products, Asp-Gly and the rearranged IsoAsp-Gly sequence, through a succinimide intermediate (Fig. 10) (Wright, 1991). This reaction provides a potential, novel means of backbone rearrangement in addition to amide cleavage for deactivating protein or peptide biological functions in vivo. Such processes are generally prevented from occurring rapidly by having the protein chains adopt conformations that exclude the favorable reaction distances and bond angles required for these processes (Clarke, 1987). Consequently, properly selected antibodies might restrain the side-chain amide carbonyl and the $(n + 1)$

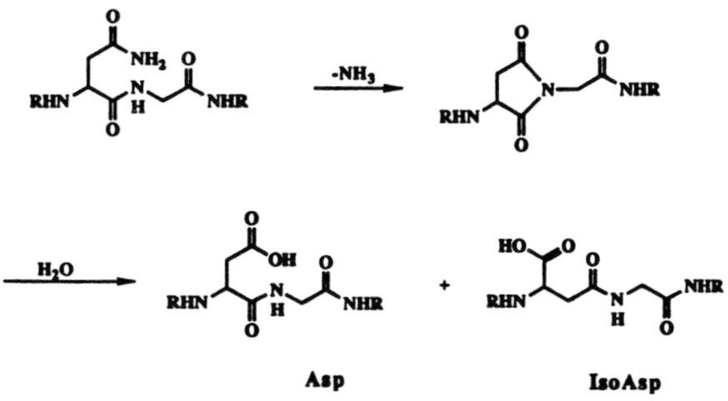

Fig. 10.   The Asn Gly reaction sequence.

amide nitrogen in an alignment and at a distance favorable for reaction to form a succinamide intermediate, which would, in turn, hydrolyze to the two products above. In addition to providing a favorable ground state conformation, the antibody binding site must also complement the intervening tetrahedral intermediates.

A model system with the corresponding hapten is shown in (Fig. 11). Some 30 monoclonal antibodies were induced and screened for their ability to catalyze both the deamidation (cyclization) and the hydrolysis (ring opening) of the succinimide (Gibbs et al., 1992b). Two antibodies, 2E4 and 24C3, had activity significantly above the background rate, as measured by either coupled glutamate dehydrogenase assay for ammonia liberation or by HPLC determination of both peptide products formed from the D-isomer as a function of time. There was no significant accumulation of the succinimide. Importantly, addition of hapten effectively blocks the catalytic process. The rate acceleration (the ratio of k_{cat} to the spontaneous reaction) for the deamidation reaction is about 100-fold, characteristic of intramolecular processes with imposed rotational restrictions (Page and Jencks, 1971). Finally, examination of the Iso/Asp product ratio for either the D- or L-isomer of the substrate showed a marked deviation from the value of 3.6 found for the spontaneous reaction. Values of IsoAsp/Asp were for the L-isomer, 2.4; for the D-isomer, 4.6 for 2E4; and 2.1 and 4.8, respectively, for 24C3, in accord with either antibody catalyzing the ring opening of the succinimide. Consequently the antibodies are catalyzing both steps of the overall reaction.

VI. CONCLUSION

The activities and stereospecificities of catalytic antibodies may result, on one hand, from conformation restrictions imposed on the ground states of

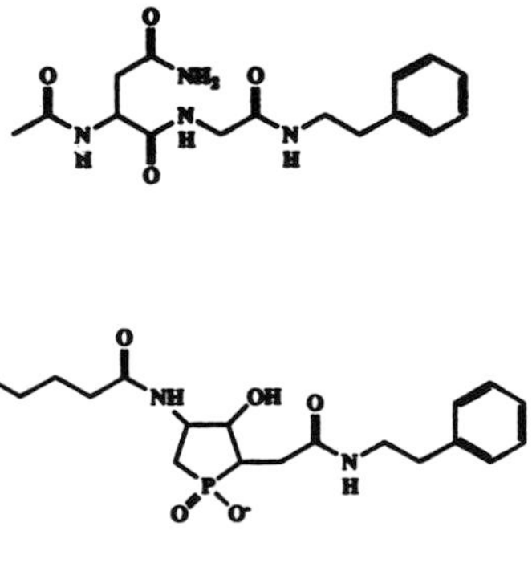

Fig. 11.   The Asn Gly model system and associated hapten.

reactants by the topological characteristics of the binding site, and, on the other hand, from more complex multistep substrate processing when the side chains of the amino acids lining the antibody binding site actively participate in the catalytic process. As anticipated, such catalytic antibodies are exquisitely sensitive to site-specific mutagenesis, which, in the case of 43C9, reinforces the importance of specific residues in the stabilization of metastable intermediates encountered along the reaction pathway.

In hindsight the opportunistic sculpting of the combining site by the immunogen provides numerous chances for creation of active sites with the kinetic and mechanistic features found in enzymes. The advent of powerful new methodologies to explore more thoroughly the immunological repertoire promises more delightful surprises to come.

REFERENCES

Benkovic, S.J., Adams, J.A., Borders, C.L., Janda, K.D., and Lerner, R.A. (1990). *Science* **250,** 1135–1139.

Benkovic, S.J., Napper, A.D., and Lerner, R.A. (1988). *Proc. Natl. Acad. Sci. U.S.A.* **85,** 5355–5358.

Better, M., Chang, C.P., Robinson, R., and Horwitz, A.H. (1988). *Science* **240,** 1041–1043.

Braisted, A., and Schultz, P.G. (1990). *J. Am. Chem. Soc.* **112,** 7430–7431.

Bruice, T.C., and Mayahi, M.F. (1960). *J. Am. Chem. Soc.* **82,** 3067–3071.

Chothia, C., Lesk, A.M., Tramontano, A., Levitt, M., Smith-Gill, S.J., Air, G., Sheriff, S., Padlan, G.A., Davies, D., Tulip, W.R., Colman, P.M., Spinelli, S., Alzari, P.M., and Poljak, R.J. (1989). *Nature* **342,** 877–883.

Clarke, S. (1987). *Int. J. Pept. Protein Res.* **30,** 808–821.

Davies, D.R., Padlan, E.A., and Sheriff, S. (1990). *Annu. Rev. Biochem.* **59,** 439–473.

Edelman, G.M., Cunningham, B.A., Gall, W.E., Gottleib, P.D., Rutishauser, U., and Waxdal, M.J. (1969). *Proc. Natl. Acad. Sci. U.S.A.* **63,** 78–85.

Getzoff, E.D., Tainer, J.A., Lerner, R.A., and Geysen, H.M. (1988). *Adv. Immunol.* **43,** 1–98.

Gibbs, R.A., Posner, B.A., Filpula, D.R., Dodd, S.W., Finkelman, M.A.J., Lee, T.K., Wroble, M., Whitlow, M., and Benkovic, S.J. (1991). *Proc. Natl. Acad. Sci. U.S.A.* **88,** 4001–4004.

Gibbs, R.A., Benkovic, P.A., Janda, K.D., Lerner, R.A., and Benkovic, S.J. (1992a). *J. Am. Chem. Soc.* **114,** 3528–3534.

Gibbs, R.A., Taylor, S., and Benkovic, S.J. (1992b). *Science* **258,** 803–805.

Hilvert, D.H., Hill, K.W., Nared, K.D., Auditor, M.T.M. (1989). *J. Am. Chem. Soc.* **111,** 9261–9262.

Huse, W.D., Sastry, L., Iverson, S.A., Kang, A.S., Alting-Mees, M., Burton, D.R., Benkovic, S.J., Lerner, R.A. (1989a). *Science* **246,** 1275–1281.

Janda, K., Schloeder, D., Benkovic, S.J., and Lerner, R.A. (1988). *Science* **241,** 1188–1191.

Janda, K.D., Weinhouse, M.I., Schloeder, D.M., Lerner, R.A., and Benkovic, S.J. (1990). *J. Am. Chem. Soc.* **112,** 1274–1275.

Jencks, W.P. (1969). "Catalysis in Chemistry and Enzymology," p. 288. McGraw- Hill, New York.

Kabat, E.A., Wu, T.T., Bilofsky, H., Reid-Miller, M., and Perry, H. (1983). "Sequences of Protein of Immunological Interest." U.S. Department of Health and Human Services, N.I.H., Washington, DC.

Kirsch, J.F., Clewell, W., and Simon, A. (1968). *J. Org. Chem.* **33**, 127–132.

Lerner, R.A., and Benkovic, S.J. (1988). *Bioessays* **9**, 107–122.

Lerner, R.A., Benkovic, S.J., and Schultz, P.G. (1991). *Science* **252**, 659–667.

Napper, A.D., Benkovic, S.J., Tramontano, A., and Lerner, R.A. (1987). *Science* **237**, 1041–1043.

Page, M.I., and Jencks, W.P. (1971). *Proc. Natl. Acad. Sci. U.S.A.* **68**, 1678–1683.

Pauling, L. (1948). *Am. Sci.* **36**, 51–58.

Pauling, L. (1990). *J. Natl. Insts. Health Res.* **2**, 59–64.

Pollack, S.J., Jacobs, J.W., and Schultz, P.G. (1986). *Science* **234**, 1570–1573.

Pollack, S.J., Nakayama, G.R., and Schultz, P.G. (1988). *Science* **242**, 1038–1040.

Posner, B., Lee, I., Itoh, T., LaPolla, R., and Benkovic, S.J. (1992). *Gene* (submitted).

Raso, V., and Stollar, B.D. (1975) *Biochemistry* **14**, 591–599.

Roberts, S., Cheetham, J.C., and Rees, A.R. (1987). *Nature* **328**, 731–734.

Sauer, J., and Sustman, R. (1980). *Angew. Chem. Int. Ed. Engl.* **19**, 779–807.

Shokat, K.M., and Schultz, P.G. (1991). *Catalytic Antibodies, Ciba Foundation Symposium* **159**, 118–134.

Skerra, A., and Pluckthun, A. (1988). *Science* **240**, 1038–1041.

Slobin, L.I. (1966). *Biochemistry* **5**, 2836–2844.

Tramontano, A., Janda, K.D., and Lerner, R.A. (1986). *Science* **234**, 1566–1570.

Watt, R.M., Herron, J.N., and Voss, E.W. (1980). *Mol. Immunol.* **17**, 1237–1243.

Wright, H.T. (1991). *CRC Revs.* **26**, 1–52.

MOLECULAR MECHANISM OF VISUAL EXCITATION

LUBERT STRYER

Department of Cell Biology, Stanford University School of Medicine,
Stanford, California

I. Introduction

IT is a pleasure and privilege for me to present a Harvey Lecture. I first savored the inviting maroon-bound volumes when I was a medical student. They opened my eyes to new realms of science and medicine, and excited my imagination. I am happy to partake in this tradition of scientific discourse, to repay in small measure the debt I owe.

The first Harvey Lecture on vision was Selig Hecht's "The Nature of the Visual Process" in 1937 (1). His bold and fundamental approach was explicitly stated: "the ultimate place of origin of the impulses which pass along the optic tracts is in the initial action of light on the receptor cells in the sense organ. Therefore, for a number of years, we have measured the different functions of vision and photoreception in man and other animals to ascertain how their quantitative properties depend on the characteristics of these very first reactions which must take place between light and the sensitive elements." The precise psychophysical experiments carried out several years later by Hecht, Shlaer, and Pirenne (2) revealed that *a retinal rod cell can be excited by a single photon.* This discovery of the exquisite sensitivity of the rod cell was profoundly influential and inspiring. I recall my sense of awe when I first read their paper.

George Wald spoke on "The Chemical Evolution of Vision" in a perceptive and prescient Harvey Lecture in 1945 (3). He began with the assertion that "chemical structure and origins provide the stuff of homologies. As such they open the possibility of pursuing biological relationship in the realm of molecular dimensions. In this sense one can speak of chemical evolution." Wald recognized nearly a half-century ago that "the processes which govern stimulation by light in all types of organisms—animal and plant—have

The Harvey Lectures, Series 87, pages 129–143
© 1993 Wiley-Liss, Inc.

many characteristics in common.'' His pioneering studies of carotenoids led a few years later to the discovery of the primary event in vision: *light triggers visual excitation by isomerizing the 11-cis retinal chromophore of visual pigments to the all-trans form* (4).

My interest in vision was also aroused by electrophysiological studies carried out by Tsuneo Tomita (5) and William Hagins (6). They found that *light hyperpolarizes retinal rod cells by transiently decreasing the large dark current of sodium ion*. This process is highly amplified: a single photon blocks the entry of millions of sodium ions. Denis Baylor and coworkers then elegantly detected the electrical response of a rod cell to a single photon (7,8). Their noise analyses showed that excitation of a single rhodopsin leads to the closure of several hundred channels in the plasma membrane.

The stage was set for biochemical studies of the molecular mechanism of visual excitation. The system was all the more alluring because much of the transduction process takes place in a discrete part of the cell which can be detached and readily purified. The outer segment contains the transduction machinery in abundant and highly concentrated form, separate from the energy-generating and protein-synthesizing apparatus. It is a gift of nature to all who care about how sensory stimuli are transduced and amplified. Indeed, a coherent picture of the molecular basis of visual excitation has recently come into view as a result of biochemical, biophysical, electrophysiological, and molecular genetic experiments carried out in many laboratories around the world. I have had the good fortune of participating in this intellectual ferment.

In essence, light triggers a nerve signal by activating an enzymatic cascade that leads to the closure of membrane channels (Fig. 1). This cascade is set in motion by the photoisomerization of the retinal chromophore of rhodopsin (R). Photoexcited rhodopsin (R*) then activates transducin (T), a member of the G-protein family, by catalyzing the exchange of GTP for bound GDP. The GTP form of transducin (specifically, $T_\alpha GTP$, the activated α subunit) then switches on a potent phosphodiesterase (PDE*) that hydrolyzes cyclic GMP (cGMP). In the dark, Na^+ and Ca^{2+} enter outer segments through cation-specific channels that are opened by the binding of cGMP. The light-triggered hydrolysis of cGMP closes channels, which hyperpolarizes the membrane and generates a neural signal. Recovery of the dark state is mediated by deactivation of PDE and activation of guanylate cyclase. The light-induced lowering of the cytosolic calcium level (Ca_i) is detected by recoverin, a recently discovered member of the EF-hand superfamily of

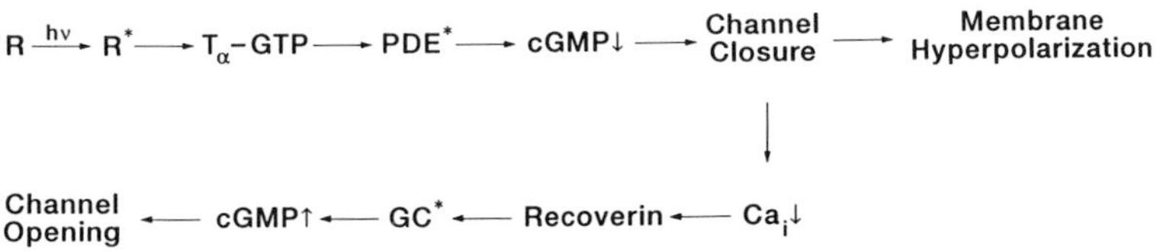

$$R \xrightarrow{h\nu} R^* \longrightarrow T_\alpha\text{-}GTP \longrightarrow PDE^* \longrightarrow cGMP\downarrow \longrightarrow \begin{array}{c}\text{Channel}\\\text{Closure}\end{array} \longrightarrow \begin{array}{c}\text{Membrane}\\\text{Hyperpolarization}\end{array}$$

$$\begin{array}{c}\text{Channel}\\\text{Opening}\end{array} \longleftarrow cGMP\uparrow \longleftarrow GC^* \longleftarrow \text{Recoverin} \longleftarrow Ca_i\downarrow$$

Fig. 1.   Flow of information in visual excitation and recovery.

calcium-binding proteins. The consequent stimulation of guanylate cyclase elevates the cGMP level, which reopens channels and restores the dark state.

II. TRANSDUCIN

My research on vision began in 1969 and was focused for nearly a decade on rhodopsin. The excitation pathway was a total enigma. The prevailing hypothesis that calcium ion serves as the excitatory transmitter bore no fruit. The first breakthrough came in 1971, when Mark Bitensky and William Miller found that light markedly lowers the level of cyclic nucleotides in rod outer segments (9). Subsequent studies showed that the reduction was due to the hydrolysis of cGMP by a specific phosphodiesterase. Interest in cyclic GMP as a transmitter candidate was greatly increased by Paul Liebman's report at a Gordon Conference in 1978 that a single photon can trigger the activation of hundreds of molecules of PDE if GTP is present (10). On the way home, I had the good fortune of staying overnight at Miller's home near Yale. After a delicious dinner, Miller showed me some intriguing electrophysiological records. He and Grant Nicol had found that the injection of cGMP into retinal rod cells led to their depolarization (11,12). Moreover, the injected nucleotide lengthened the delay between the arrival of a light pulse and membrane hyperpolarization. The simplest interpretation was that cGMP opened channels, which stayed open until cGMP was degraded by a light-activated PDE.

When I returned to Stanford, I discussed these intriguing findings with Bernard Fung, a postdoctoral fellow in my laboratory. We now viewed cGMP as an attractive transmitter candidate and decided to explore the molecular mechanism of the light-triggered activation of PDE. The key clues that guided our design of experiments were: 1) GTP in addition to light is needed to activate the enzyme, and 2) rod outer segments contain a light-activated GTPase. We surmised that a guanyl-nucleotide binding protein plays a vital role in the switching of PDE. Our first experiments were therefore designed to detect the binding of GTP to rod outer segment membranes (ROS). Mem-

branes were incubated in the dark with [α-^{32}P]GTP and washed with a
nucleotide-free buffer. We were rewarded by finding that radioactive GDP
was tightly bound to the ROS, and interestingly, that illumination led to the
release of GDP (13). A striking effect was then observed: GTP markedly
enhanced the action of light in releasing bound GDP. In the presence of 1
μM GTP, release was half-maximal when fewer than 1 in 10,000 rhodopsins
were excited.

The effect of light on the uptake of GTP was then studied. GppNHp, a
hydrolysis-resistant analog of GTP, was used to simplify the analysis. We
found that the uptake of GppNHp, like the release of GDP, requires the excitation of only a small fraction of the rhodopsin molecules (13). Furthermore,
the kinetics of GppNHp uptake paralleled that of GDP release. Most significant, a single R* led to the uptake of 500 molecules of GppNHp (Fig. 2).
These experiments revealed that *photoexcited rhodopsin catalyzes the exchange of GTP for bound GDP in a rod outer segment protein.*

Our observation of a highly amplified GTP-GDP exchange led us to pro-

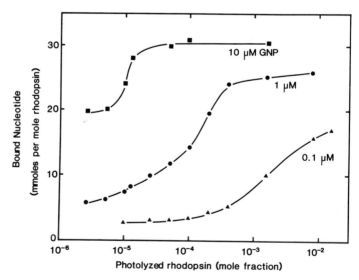

Fig. 2.   Photoexcited rhodopsin catalyzes the entry of a GTP analog into transducin.
The dependence of the binding of GppNHp by ROS membranes on the mole fraction
of R in 0.1 μM. (▲), 1 μM (●), and 10 μM (■) GppNHp is shown. Reproduced from
Fung and Stryer (13), with permission of the publisher.

pose that a guanyl nucleotide binding protein, now called transducin (T), serves as the information-carrying intermediate in the light-triggered activation of the phosphodiesterase.

$$R^* \rightarrow \text{T-GTP} \rightarrow \text{PDE}^*$$

This hypothesis predicted that 1) T-GTP can be formed in the absence of the phosphodiesterase, and 2) the phosphodiesterase can be activated by T-GTP in the absence of R^*.

The next step was to purify transducin, a task made easier by Hermann Kühn's key finding that T-GDP binds tightly to disk membranes, whereas T-GTP does not (14). We found that transducin consists of three polypeptide chains: α (39 kd), β (36 kd), and γ (8 kd). The availability of purified transducin enabled us to test the first prediction of the proposed flow of information. Purified transducin was incubated in the dark with reconstituted membranes consisting of phosphatidylcholine and rhodopsin. A striking result was obtained: a single R^* catalyzed the uptake of 71 molecules of GppNHp in this reconstituted system (15). This was the first time that an amplified effect of rhodopsin requiring only one other protein—transducin—had been found.

III. PHOSPHODIESTERASE

We then proceeded to test the second prediction of our proposed flow of information. James Hurley joined my laboratory and participated in determining whether the GTP form of transducin can activate the phosphodiesterase in the dark. The first step was to purify the GppNHp form of transducin. When we did so, we encountered a surprise. The high-pressure liquid chromatography (HPLC) elution profile showed that transducin had dissociated into T_α and $T_{\beta\gamma}$ subunits following the exchange of GppNHp for GDP. This made it possible to determine which subunit contains the guanyl-nucleotide binding site. All of the radioactivity eluted in the α subunit, showing that T_α contains the binding site for GTP or GDP. Does T_α-GppNHp or $T_{\beta\gamma}$ activate PDE when added to unilluminated disk membranes? Our experiment (Fig. 3) provided a decisive answer: T_α containing the GTP analog fully activates the phosphodiesterase, whereas $T_{\beta\gamma}$ does not (15). These experiments established that *T_α-GTP is the information-carrying intermediate in the light-triggered activation of the phosphodiesterase.*

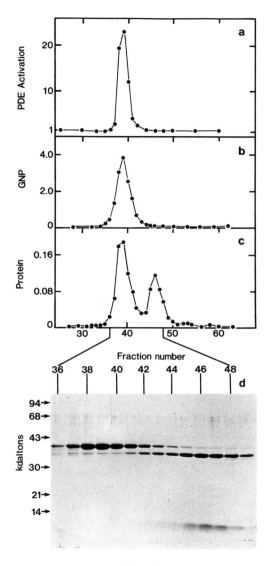

Figure 3.

How does T_α-GTP activate PDE? A valuable clue was provided by Bitensky's finding that PDE can be activated by the addition of a small amount of trypsin, suggesting that the enzyme is subject to an inhibitory constraint in the dark. Hurley then found that the small γ subunit (9 kd) of the PDE is rapidly degraded by PDE, whereas the α (88 kd) and β (85 kd) subunits are left largely intact (16). The holoenzyme has the subunit structure $\alpha\beta\gamma_2$ (17). The trypsin-activated $\alpha\beta$ enzyme is highly active. Furthermore, its catalytic activity can be nearly completely switched off by adding purified γ subunit, which binds very tightly ($K_d = 10$ pM) (16). Thus, *the phosphodiesterase consists of separable regulatory and catalytic subunits.* Site-specific mutagenesis studies of recombinant γ subunit have shown that the carboxyl terminus of this inhibitory subunit is essential for blocking catalytic activity (18,19). The maximal activity obtained by activating PDE with transducin is nearly the same as that achieved by proteolytic destruction of the γ subunit. Thus, *T_α-GTP activates PDE by relieving the inhibitory constraint imposed by its γ subunits.* The catalytic prowess of activated PDE is impressive; its K_{cat}/K_m ratio of 6×10^7 M^{-1} s^{-1} is near the limit set by the diffusion-controlled encounter of enzyme and substrate.

A light-activated amplification cycle involving rhodopsin, transducin, and PDE is shown in Figure 4. In the dark, nearly all of the transducin is in the T-GDP form, which does not activate PDE. R* encounters T-GDP by lateral diffusion in the plane of the membrane. An R*-T-GDP complex is formed. GDP is released, leaving an empty guanyl-nucleotide binding site. GTP then binds to form an R*-T-GTP complex, which rapidly dissociates. R* is released to catalyze another round of activation of transducin, and Tα-GTP carries the excitation signal forward to PDE. Activated PDE then hydrolyzes cGMP at a rapid rate. The GTPase activity inherent in the α subunit of transducin converts T_α-GTP into T_α-GDP, thereby deactivating the PDE. R* must also

Fig. 3.   The α-subunit of transducin bearing GTP is the activator of the cGMP phosphodiesterase. Transducin containing a bound GTP analog (GppNHp) was applied to a high-pressure liquid chromatography column. The elution profile and characteristics of the fractions are shown here. **a:** Stimulation of phosphodiesterase activity of unilluminated ROS membranes by addition of an aliquot of each fraction. **b:** Distribution of radioactive GppNHp, showing that the α-subunit of transducin contains the bound nucleotide. **c:** Protein profile, showing that the α and $\alpha\gamma$ subunits of transducin are dissociated when the protein is in the GTP form. Reproduced from Fung, Hurley, and Stryer (15), with permission of the publisher.

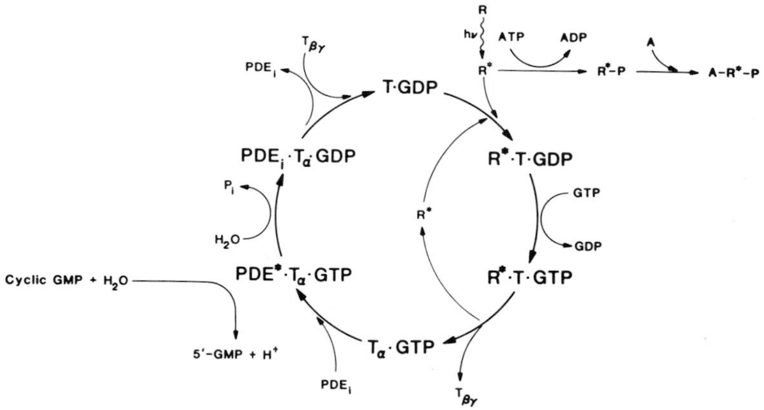

Fig. 4.   Light-activated transducin cycle. R, unexcited rhodopsin; R*, photoexcited rhodopsin; R*-P, multiply phosphorylated, photoexcited rhodopsin; T, transducin; PDE_i, inhibited phosphodiesterase; PDE*, activated phosphodiesterase; A, arrestin.

be deactivated to prevent continuing activation of transducin. This is accomplished by a kinase that phosphorylates multiple serines and threonines in the carboxy-terminal tail of R* (20,21). The capping of phosphorylated R* by arrestin totally switches off its capacity to activate transducin.

How fast is the activation of transducin? Minh Vuong, a graduate student in my laboratory, was joined by Marc Chabre, who came on a sabbatical visit. They carried out infrared light-scattering experiments on magnetically oriented rod outer segments. A highly anisotropic signal arising from the release of T_α-GTP from the membrane served as a window on the kinetics of activation. We found that a molecule of R* activates a molecule of transducin in 1 ms (22). Thus, several hundred transducins can be activated on the subsecond time scale, which is sufficiently rapid to account for the kinetics of the electrical response of a rod to a single photon (7).

IV. CYCLIC-GMP-GATED CHANNEL

How does the light-triggered hydrolysis of cGMP lead to a nerve signal? Evginiy Fesenko's incisive patch-clamp studies of excised patches of the rod outer segment plasma membrane provided a surprisingly simple answer: *cytosolic cGMP directly opens cation-specific channels, whereas calcium ion and nucleoside triphosphates have no effect* (23). Channel opening is highly cooperative (the Hill coefficient is 3), which renders it highly responsive to small changes in the cGMP level. The channel, an oligomer of 80-kd subunits, has been functionally reconstituted in lipid bilayer membranes (24,25).

I have been fortunate in having a close association with Denis Baylor, who has enriched my understanding of membrane channels and electrophysiology. Our laboratories have collaborated in several studies of the cGMP-gated channel. We found that the introduction of 8-bromo-cGMP, a hydrolysis-resistant analog of cGMP, into intact retinal rod cells blocks the normal light-induced decrease in membrane current, strongly suggesting that the channel is gated solely by cGMP (26). Under physiological conditions, the channel opens and closes in times of milliseconds in response to changes in the cGMP level, as shown by photolysis of a caged analog of cGMP (Fig. 5) and voltage-jump conductance studies (27,28). Thus, *the plasma membrane of the rod outer segment is in essence a cGMP electrode. The channel responds to the instantaneous level of cGMP, which it samples each millisecond.*

V. Guanylate Cyclase and Recoverin

Restoration of the dark state depends on the deactivation of the phosphodiesterase and the resynthesis of cGMP. What is the signal for activation of guanylate cyclase? The finding that the cGMP content of ROS increases several-fold when Ca_i is markedly lowered pointed to a negative feedback loop between cGMP and Ca_i (29). In the dark, the entry of Ca^{2+} through the cGMP-gated channel is matched by its efflux through an exchanger that is driven by the influx of three Na^+ and the efflux of one K^+ (30,31). Karl Koch and I then found that the activity of guanylate cyclase is exquisitely sensitive to the calcium level in the nanomolar range (32). Cyclase activity

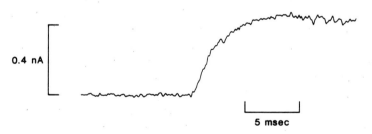

0.4 nA

5 msec

Fig. 5.   Millisecond opening of the cyclic-GMP-gated channel following the uncaging of a cGMP analog by a laser pulse. Reproduced from Karpen, Zimmerman, Stryer, and Baylor (27), with permission of the publisher.

increased more than fivefold when Ca_i was lowered to less than about 100 nM (Fig. 6). The steep dependence of cyclase on Ca_i indicates that Ca^{2+} acts cooperatively. Stimulation at low Ca_i was shown to depend on a protein that can be detached from the membrane with a low ionic strength buffer. *Thus, guanylate cyclase, like the phosphodiesterase, consists of separable regulatory and catalytic subunits.*

Mitra Ray, a graduate student in my laboratory, obtained a soluble fraction that activated guanylate cyclase in a calcium-sensitive manner. The fraction contained two calcium-binding proteins with masses in the 20 to 30 kd range. She discussed her findings with James Hurley, now a professor at the University of Washington. Hurley told her that Sasha Dizhoor brought from Moscow a 23-kd calcium-binding protein from rod outer segments that he had recently isolated. Ken Walsh had sequenced most of the protein. Also, Hurley's laboratory had prepared antibodies to it and found that the protein was localized in retinal rod and cone cells. Ray then assayed this protein and found that it activated cyclase when the calcium level was lowered (33). We named this protein *recoverin* because it participates in restoring the dark state. Recoverin was independently isolated at the same time by Koch and Lambrecht (34). The amino acid sequence of recoverin revealed that it contains multiple EF hands of the kind seen in calmodulin and other members of this superfamily of calcium-binding proteins.

Sergey Zozulya joined my laboratory at this time and focused his efforts on recoverin in a continuation of our collaboration with Hurley's group. Hundreds of milligrams of purified recoverin were obtained by expressing the cDNA in *E. coli* (35). We were initially puzzled by the observation that the recombinant protein differed from the retinal one in its isoelectric point and in the effect of calcium on its tryptophan fluorescence emission spectrum. Mass spectrometry (36) then revealed that the amino terminus of retinal recoverin is acylated with a myristoyl (14:0) group or a related fatty acyl chain (12:0, 14:1, or 14:2). We then prepared myristoylated recombinant recoverin by coexpressing the cDNAs for recoverin and yeast *N*-myristoyltransferase in *E. coli*. The myristoylated recombinant protein is like the native retinal protein. What is the role of myristoylation? An intriguing finding is that Ca^{2+} induces the insertion of native recoverin and myristoylated recombinant recoverin, but not unmyristoylated recombinant recoverin, into lipid bilayer membranes. *Recoverin contains a calcium-myristoyl switch that probably plays a key role in transducing changes in calcium concentration into changes in the activity of guanylate cyclase.* One of our aims now is to

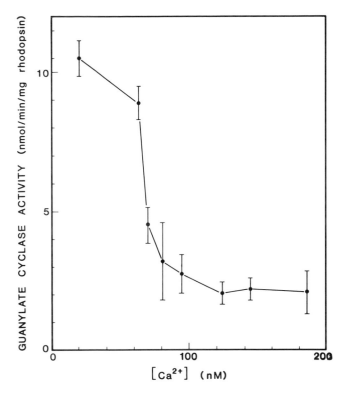

Fig. 6.   Calcium dependence of the catalytic activity of guanylate cyclase. Reproduced from Koch and Stryer (32), with permission of the publisher.

delineate the mechanism of activation and reconstitute calcium-sensitive activation of cyclase in a well-defined model system, as has been accomplished for transducin and the phosphodiesterase.

The availability of large amounts of unmyristoylated recombinant recoverin has opened the door to high-resolution X-ray crystallographic and nuclear magnetic resonance studies of this new calcium sensor. Zozulya crystallized calcium-bound unmyristoylated recoverin and gave the crystals to David McKay and Kevin Flaherty, who solved the structure in only a few months. The protein contains four EF hands, two of which can bind calcium (or a lanthanide such as samarium). The four EF hands are tandemly arranged to form a compact molecule, in contrast with calmodulin and troponin C, where pairs of

EF hands are separated from one another by a long helix to give a dumbbell shape (37). Nature combines EF-hand modules in many different ways to generate diverse calcium sensors (38). We would like to crystallize and solve the structure of the calcium-free form of unmyristoylated recoverin, and of both forms of the myristoylated protein, to learn how the calcium-myristoyl switch operates at the atomic level.

VI. RECURRING MOTIFS

Hecht's research a half-century ago defined a major question in the field of vision: How does a single photon trigger a retinal rod cell? We now know the molecular pathway in outline (Fig. 7). The photoexcitation of rhodopsin activates transducin, which in turn switches on a phosphodiesterase. The consequent decrease in the cGMP level closes channels to hyperpolarize the membrane and generate a neural signal. Excitation automatically sets in motion the recovery process by lowering the cytosolic calcium level. The acti-

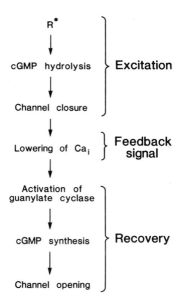

Fig. 7.   Excitation of retinal rods sets in motion the recovery process. The light-induced lowering of the cytosolic calcium level is the feedback signal that leads to the restoration of the dark state.

vation of guanylate cyclase leads to the resynthesis of cGMP, which once again opens channels. Recoverin participates in the recovery process by serving as a calcium sensor. Thus, vision involves the interplay of two messengers. *Cyclic GMP carries the excitation signal, whereas calcium ion is the remembrance of photons past.* The challenge now is to elucidate the molecular mechanism of adaptation, the remarkable process that enables us to perceive contrast over a hundred-thousand-fold range of background light level.

The other major question—How did vision evolve?—was posed decades ago by Wald. Molecular genetic studies have revealed that rod and cone visual pigments are variations on the same molecular theme (39). Furthermore, the transducin, phosphodiesterase, and cGMP-gated channel proteins of cones are very much like their counterparts in rods. Indeed, rod and cone transduction mechanisms are very similar. Moreover, the initial events in visual excitation in invertebrates closely resemble those of vertebrates. In the same vein, olfaction begins with a cyclic nucleotide cascade resembling that of vision (40,41). The seven-helix rhodopsin motif is found in many other signal transduction systems, as exemplified by adrenergic receptors in hormone action (42). This motif is an early eukaryotic invention, as illustrated by yeast mating factor receptors (43). Transducin is the best understood of the G-proteins, the large family of signal-coupling proteins that play key roles in sensory transduction, hormone action, and growth (44). And now, with recoverin, we have an evolutionary tie between vision and other calcium-sensitive processes.

The study of the molecular mechanism of vision has been immensely rewarding in revealing recurring motifs of signal transduction. We are also catching glimpses of how this beautiful process came into being.

ACKNOWLEDGMENTS

I am grateful to my colleagues and collaborators for many wonderful adventures and explorations of vision. My research on vision has been supported by the National Eye Institute and the National Institute of General Medical Sciences.

REFERENCES

1. Hecht, S. (1938). *Harvey Lect.* **33,** 35–64.
2. Hecht, S., Shlaer, S., and Pirenne, M.H. (1942). *J. Gen. Physiol.* **25,** 819–840.
3. Wald, G. (1946). *Harvey Lect.* **41,** 117–160.
4. Wald, G. (1968). *Nature* **219,** 800–807.
5. Tomita, T. (1970). *Q. Rev. Biophys.* **3,** 179–222.

142 LUBERT STRYER

6. Hagins, W.A., Penn, R.D., and Yoshikami, S. (1970). *Biophys. J.* **10**, 380–412.
7. Baylor, D.A., Lamb, T.D., and Yau, K.W. (1979). *J. Physiol. (Lond).* **288**, 613–634.
8. Baylor, D.A. (1987). *Invest. Ophthalmol. Vis. Sci.* **28**, 34–49.
9. Bitensky, M.W., Gorman, R.E. and Miller, W.H. (1971). *Proc. Natl. Acad. Sci. U.S.A.* **68**, 561–562.
10. Yee, R. and Liebman, P.A. (1978). *J. Biol. Chem.* **253**, 8902–8909.
11. Miller, W.H. and Nicol, G.D. (1979). *Nature* **280**, 64–66.
12. Miller, W.H. (1990). *Invest. Ophthalmol. Vis. Sci.* **31**, 1664–1673.
13. Fung, B.K.-K., and Stryer, L. (1980). *Proc. Natl. Acad. Sci. U.S.A.* **77**, 2500–2504.
14. Kühn, H. (1980). *Nature* **283**, 587–589.
15. Fung, B.K.-K., Hurley, J.B., and Stryer, L. (1981). *Proc. Natl. Acad. Sci. U.S.A.* **78**, 152–156.
16. Hurley, J.B., and Stryer, L. (1982). *J. Biol. Chem.* **257**, 11094–11099.
17. Deterre, P., Bigay, J., Robert, M., Pfister, C., Kuhn, H., and Chabre, M. (1986). *Proteins Struct. Funct. Genet.* **1**, 188–193.
18. Brown, R.L., and Stryer, L. (1989). *Proc. Natl. Acad. Sci. U.S.A.* **86**, 4922–4926.
19. Brown, R.L. (1992). *Biochemistry* **31**, 5918–5925.
20. Wilden, U., Hall, S.W., and Kühn, H. (1986). *Proc. Natl. Acad. Sci. U.S.A.* **83**, 1174–1178.
21. Miller, J.L., Fox, D.A., and Litman, B.J. (1986). *Biochemistry* **25**, 4983–4988.
22. Vuong, T.M., Chabre, M., and Stryer, L. (1984). *Nature* **311**, 659–661.
23. Fesenko, E.E., Kolesnikov, S.S., and Lyubarsky, A.L. (1985). *Nature* **313**, 310–313.
24. Kaupp, U.B., Hanke, W., Simmoteit, R., and Luhring, H. (1988). *Cold Spring Harb. Symp. Quant. Biol.* **53**, 407–415.
25. Kaupp, U.B., Niidome, T., Tanabe, T., Terade, S., Bonigk, W., Stuhmer, W., Cook, N.J., Kangawa, K., Matsuo, H., Hirose, T., Miyata, T., and Numa, S. (1989). *Nature* **342**, 762–766.
26. Zimmerman, A.L., Yamanaka, G., Eckstein, F., Baylor, D.A., and Stryer, L. (1985). *Proc. Natl. Acad. Sci. U.S.A.* **82**, 8813–8817.
27. Karpen, J.W., Zimmerman, A.L., Stryer, L., and Baylor, D.A. (1988). *Proc. Natl. Acad. Sci. U.S.A.* **85**, 1287–1291.
28. Karpen, J.W., Zimmerman, A.L., Stryer, L., and Baylor, D.A. (1988). *Cold Spring Harb. Symp. Quant. Biol.* **53**, 325–332.
29. Lipton S.A., and Dowling, J.E. (1981). *Curr. Top. Membr. Transp.* **15**, 381–392.
30. Yau, K.W., and Nakatani, K. (1985). *Nature* **313**, 579–582.
31. Cervetto, L., Lagnado, L., Perry, R.J., Robinson, D.W., and McNaughton, P.A. (1989). *Nature* **337**, 740–743.
32. Koch, K.W., and Stryer, L. (1988). *Nature* **344**, 64–66.
33. Dizhoor, A.M., Ray, S., Kumar, S., Niemi, G., Spencer, M., Brolley, D., Walsh, K.A., Philipov, P.P., Hurley, J.B., and Stryer, L. (1991). *Science* **251**, 915–918.
34. Lambrecht, H.G., and Koch, K.W. (1991). *EMBO J.* **10**, 793–798.
35. Ray, S., Zozulya, S., Niemi, G.A., Flaherty, K.M., Brolley, D., Dizhoor, A.M., McKay, D.B., Hurley, J., and Stryer, L. (1992). *Proc. Natl. Acad. Sci. U.S.A.* **89**, 5705–5709.
36. Dizhoor, A.M., Ericsson, L.H., Johnson, R.S., Kumar, S., Olshevskaya, E., Zozulya,

S., Neubert, T.A., Stryer, L., Hurley, J.B., and Walsh, K.A. (1992). *J. Biol. Chem.* **267,** 16033–16036.

37. Strynadka, N.C.J., and James, M.N.G. (1989). *Annu. Rev. Biochem.* **58,** 951–998.
38. Kretsinger, R.H. (1987). *Cold Spring Harbor Symp. Quant. Biol.* **52,** 499–510.
39. Nathans, J., Thomas, D., and Hogness, D.S. (1986). *Science* **232,** 193–202.
40. Lancet, D., Lazard, D., Heldman, J., Khen, M., and Nef, P. (1988). *Cold Spring Harb. Symp. Quant. Biol.* **53,** 343–348.
41. Buck, L., and Axel, R. (1991). *Cell* **65,** 175–187.
42. Hargrave, P.A. (1991). *Curr. Opin. Struct. Biol.* **1,** 575–581.
43. Blumer, K.J., Reneke, J.E., Courchesne, W.E., and Thorner, J. (1988). *Cold Spring Harb. Symp. Quant. Biol.* **53,** 591–603.
44. Bourne, H.R., Sanders, D.A., and McCormick, F. (1990). *Nature* **348,** 125–132.

FORMER OFFICERS OF THE HARVEY SOCIETY

---

* At the Annual Meeting of May 18, 1909, these officers were elected. In publishing the 1909–1910 volume their names were omitted, possibly because in that volume the custom of publishing the names of the incumbents of the current year was changed to publishing the names of the officers selected for the ensuing year.

1911–1912

President: S.J. MELTZER
Vice-President: FREDERIC S. LEE
Treasurer: EDWARD K. DUNHAM
Secretary: HAVEN EMERSON

Council:
GRAHAM LUSK
JAMES EWING
SIMON FLEXNER

1912–1913

President: FREDERIC S. LEE
Vice-President: WM. H. PARK
Treasurer: EDWARD K. DUNHAM
Secretary: HAVEN EMERSON

Council:
GRAHAM LUSK
S.J. MELTZER
WM. G. MACCALLUM

1913–1914

President: FREDERIC S. LEE
Vice-President: WM. G. MACCALLUM
Treasurer: EDWARD K. DUNHAM
Secretary: AUGUSTUS B. WADSWORTH

Council:
GRAHAM LUSK
WM H. PARK
GEORGE B. WALLACE

1914–1915

President: WM. G. MACCALLUM
Vice-President: RUFUS I. COLE
Treasurer: EDWARD K. DUNHAM
Secretary: JOHN A. MANDEL

Council:
GRAHAM LUSK
FREDERIC S. LEE
W.T. LONGCOPE

1915–1916

President: GEORGE B. WALLACE
Treasurer: EDWARD K. DUNHAM
Secretary: ROBERT A. LAMBERT

Council:
GRAHAM LUSK
RUFUS I. COLE
NELLIS B. FOSTER

1916–1917

President: GEORGE B. WALLACE
Vice-President: RUFUS I. COLE
Treasurer: EDWARD K. DUNHAM
Secretary: ROBERT A. LAMBERT

Council:
GRAHAM LUSK†
W.T. LONGCOPE
S.R. BENEDICT
HANS ZINSSER

---

* Dr. William G. MacCallum resigned after election. On Doctor Lusk's motion Doctor George B. Wallace was made President—no Vice President was appointed.

†Doctor Lusk was made Honorary permanent Counsellor.

1917–1918

President: EDWARD K. DUNHAM
Vice-President: RUFUS I. COLE
Treasurer: F.H. PIKE
Secretary: A.M. PAPPENHEIMER

Council:
GRAHAM LUSK
GEORGE B. WALLACE
FREDERIC S. LEE
PEYTON ROUS

1918–1919

President: GRAHAM LUSK
Vice-President: RUFUS I. COLE
Treasurer: F.H. PIKE
Secretary: K.M. VOGEL

Council:
GRAHAM LUSK
JAMES W. JOBLING
FREDERIC S. LEE
JOHN AUER

1919–1920

President: WARFIELD T. LONGCOPE
Vice-President: S.R. BENEDICT
Treasurer: F.H. PIKE
Secretary: K.M. VOGEL

Council:
GRAHAM LUSK
HANS ZINSSER
FREDERIC S. LEE
GEORGE B. WALLACE

1920–1921*

President: WARFIELD T. LONGCOPE
Vice-President: S.R. BENEDICT
Treasurer: A.M. PAPPENHEIMER
Secretary: HOMER F. SWIFT

Council:
GRAHAM LUSK
FREDERIC S. LEE
HANS ZINSSER
GEORGE B. WALLACE

1921–1922

President: RUFUS I. COLE
Vice-President: S.R. BENEDICT
Treasurer: A.M. PAPPENHEIMER
Secretary: HOMER F. SWIFT

Council:
GRAHAM LUSK
HANS ZINSSER
H.C. JACKSON
W.T. LONGCOPE

1922–1923

President: RUFUS I. COLE
Vice-President: HANS ZINSSER
Treasurer: CHARLES C. LIEB
Secretary: HOMER F. SWIFT

Council:
GRAHAM LUSK
W.T. LONGCOPE
H.C. JACKSON
S.R. BENEDICT

*These officers were elected at the Annual Meeting of May 21, 1920 but were omitted in the publication of the 1919–20 volume.

1923–1924

President: EUGENE F. DUBOIS
Vice-President: HOMER F. SWIFT
Treasurer: CHARLES C. LIEB
Secretary: GEORGE M. MACKENZIE

Council:
 GRAHAM LUSK
 ALPHONSE R. DOCHEZ
 DAVID MARINE
 PEYTON ROUS

1924–1925

President: EUGENE F. DUBOIS
Vice-President: PEYTON ROUS
Treasurer: CHARLES C. LIEB
Secretary: GEORGE M. MACKENZIE

Council:
 GRAHAM LUSK
 RUFUS COLE
 HAVEN EMERSON
 WM. H. PARK

1925–1926

President: HOMER F. SWIFT
Vice-President: H.B. WILLIAMS
Treasurer: HAVEN EMERSON
Secretary: GEORGE M. MACKENZIE

Council:
 GRAHAM LUSK
 EUGENE F. DUBOIS
 WALTER W. PALMER
 H.D. SENIOR

1926–1927

President: WALTER W. PALMER
Vice-President: WM. H. PARK
Treasurer: HAVEN EMERSON
Secretary: GEORGE M. MACKENZIE

Council:
 GRAHAM LUSK
 HOMER F. SWIFT
 A.R. DOCHEZ
 ROBERT CHAMBERS

1927–1928

President: DONALD D. VAN SLYKE
Vice-President: JAMES W. JOBLING
Treasurer: HAVEN EMERSON
Secretary: CARL A.L. BINGER

Council:
 GRAHAM LUSK
 RUSSEL L. CECIL
 WARD J. MACNEAL
 DAVID MARINE

1928–1929

President: PEYTON ROUS
Vice-President: HORATIO B. WILLIAMS
Treasurer: HAVEN EMERSON
Secretary: PHILIP D. MCMASTER

Council:
 GRAHAM LUSK
 ROBERT CHAMBERS
 ALFRED F. HESS
 H.D. SENIOR

1929–1930

President: G. CANBY ROBINSON
Vice-President: ALFRED F. HESS
Treasurer: HAVEN EMERSON
Secretary: DAYTON J. EDWARDS

Council:
 GRAHAM LUSK
 ALFRED E. COHN
 A.M. PAPPENHEIMER
 H.D. SENIOR

1930–1931

President: ALFRED E. COHN
Vice-President: J.G. HOPKINS
Treasurer: HAVEN EMERSON
Secretary: DAYTON J. EDWARDS

Council:
GRAHAM LUSK
O.T. AVERY
A.M PAPPENHEIMER
S.R. DETWILER

1931–1932

President: J.W. JOBLING
Vice-President: HOMER W. SMITH
Treasurer: HAVEN EMERSON
Secretary: DAYTON J. EDWARDS

Council:
GRAHAM LUSK
S.R. DETWILER
THOMAS M. RIVERS
RANDOLPH WEST

1932–1933

President: ALFRED F. HESS
Vice-President: HAVEN EMERSON
Treasurer: THOMAS M. RIVERS
Secretary: EDGAR STILLMAN

Council:
GRAHAM LUSK
HANS T. CLARKE
WALTER W. PALMER
HOMER W. SMITH

1933–1934

President: ALFRED HESS*
Vice-President: ROBERT K. CANNAN
Treasurer: THOMAS M. RIVERS
Secretary: EDGAR STILLMAN

Council:
STANLEY R. BENEDICT
ROBERT F. LOEB
WADE H. BROWN

1934–1935

President: ROBERT K. CANNAN
Vice-President: EUGENE L. OPIE
Treasurer: THOMAS M. RIVERS
Secretary: RANDOLPH H. WEST

Council:
HERBERT S. GASSER
B.S. OPPENHEIMER
PHILIP E. SMITH

1935–1936

President: ROBERT K. CANNAN
Vice-President: EUGENE L. OPIE
Treasurer: THOMAS M. RIVERS
Secretary: RANDOLPH H. WEST

Council:
ROBERT F. LOEB
HOMER W. SMITH
DAVID MARINE

1936–1937

President: EUGENE L. OPIE
Vice-President: PHILIP E. SMITH
Treasurer: THOMAS M. RIVERS
Secretary: MCKEEN CATTELL

Council:
GEORGE B. WALLACE
MARTIN H. DAWSON
JAMES B. MURPHY

---

* Dr. Hess died December 5, 1933.

1937–1938

President: EUGENE L. OPIE
Vice-President: PHILIP E. SMITH
Treasurer: THOMAS M. RIVERS
Secretary: MCKEEN CATTELL

Council:
   GEORGE B. WALLACE
   MARTIN H. DAWSON
   HERBERT S. GASSER

1938–1939

President: PHILIP E. SMITH
Vice-President: HERBERT S. GASSER
Treasurer: KENNETH GOODNER
Secretary: MCKEEN CATTELL

Council:
   HANS T. CLARKE
   JAMES D. HARDY
   WILLIAM S. TILLETT

1939–1940

President: PHILIP E. SMITH
Vice-President: HERBERT S. GASSER
Treasurer: KENNETH GOODNER
Secretary: THOMAS FRANCIS, JR.

Council:
   HANS T. CLARKE
   N. CHANDLER FOOT
   WILLIAM S. TILLETT

1940–1941

President: HERBERT S. GASSER
Vice-President: HOMER W. SMITH
Treasurer: KENNETH GOODNER
Secretary: THOMAS FRANCIS, JR.

Council:
   N. CHANDLER FOOT
   VINCENT DU VIGNEAUD
   MICHAEL HEIDELBERGER

1941–1942

President: HERBERT S. GASSER
Vice-President: HOMER W. SMITH
Treasurer: KENNETH GOODNER
Secretary: JOSEPH C. HINSEY

Council:
   HARRY S. MUSTARD
   HAROLD G. WOLFF
   MICHAEL HEIDELBERGER

1942–1943

President: HANS T. CLARKE
Vice-President: THOMAS M. RIVERS
Treasurer: KENNETH GOODNER
Secretary: JOSEPH C. HINSEY

Council:
   ROBERT F. LOEB
   HAROLD G. WOLFF
   WILLIAM C. VON GLAHN

1943–1944

President: HANS T. CLARKE
Vice-President: THOMAS M. RIVERS
Treasurer: COLIN M. MACLEOD
Secretary: JOSEPH C. HINSEY

Council:
   ROBERT F. LOEB
   WILLIAM C. VON GLAHN
   WADE W. OLIVER

1944–1945

President: ROBERT CHAMBERS
Vice-President: VINCENT DU VIGNEAUD
Treasurer: COLIN M. MACLEOD
Secretary: JOSEPH C. HINSEY

Council:
   WADE W. OLIVER
   MICHAEL HEIDELBERGER
   PHILIP D. MCMASTER

1945–1946

President: ROBERT CHAMBERS
Vice-President: VINCENT DU VIGNEAUD
Treasurer: COLIN M. MACLEOD
Secretary: EDGAR G. MILLER, JR.

Council:
  PHILIP D. MCMASTER
  EARL T. ENGLE
  FRED W. STEWART

1946–1947

President: VINDENT DU VIGNEAUD
Vice-President: WADE W. OLIVER
Treasurer: COLIN M. MACLEOD
Secretary: EDGAR G. MILLER, JR.

Council:
  EARL T. ENGLE
  HAROLD G. WOLFF
  L. EMMETT HOLT, JR.

1947–1948

President: VINCENT DU VIGNEAUD
Vice-President: WADE W. OLIVER
Treasurer: HARRY B. VAN DYKE
Secretary: MACLYN MCCARTY

Council:
  PAUL KLEMPERER
  L. EMMETT HOLT, JR.
  HAROLD G. WOLFF

1948–1949

President: WADE W. OLIVER
Vice-President: ROBERT F. LOEB
Treasurer: HARRY B. VAN DYKE
Secretary: MACLYN MCCARTY

Council:
  PAUL KLEMPERER
  SEVERO OCHOA
  HAROLD L. TEMPLE

1949–1950

President: WADE W. OLIVER
Vice-President: ROBERT F. LOEB
Treasurer: JAMES B. HAMILTON
Secretary: MACLYN MCCARTY

Council:
  WILLIAM S. TILLETT
  SEVERO OCHOA
  HAROLD L. TEMPLE

1950–1951

President: ROBERT F. LOEB
Vice-President: MICHAEL HEIDELBERGER
Treasurer: JAMES B. HAMILTON
Secretary: LUDWIN W. EICHNA

Council:
  WILLIAM S. TILLETT
  A.M. PAPPENHEIMER, JR.
  DAVID P. BARR

1951–1952

President: RENÉ J. DUBOIS
Vice-President: MICHAEL HEIDELBERGER
Treasurer: JAMES B. HAMILTON
Secretary: LUDWIN W. EICHNA

Council:
  DAVID P. BARR
  ROBERT F. PITTS
  A.M. PAPPENHEIMER, JR.

1952–1953

President: MICHAEL HEIDELBERGER
Vice-President: SEVERO OCHOA
Treasurer: CHANDLER MCC. BROOKS
Secretary: HENRY D. LAUSON

Council:
  ROBERT F. PITTS
  JEAN OLIVER
  ALEXANDER B. GUTMAN

1953–1954

President: SEVERO OCHOA
Vice-President: DAVID P. BARR
Treasurer: CHANDLER McC. BROOKS
Secretary: HENRY D. LAUSON

Council:
  JEAN OLIVER
  ALEXANDER B. GUTMAN
  ROLLIN D. HOTCHKISS

1954–1955

President: DAVID P. BARR
Vice-President: COLIN M. MACLEOD
Treasurer: CHANDLER McC. BROOKS
Secretary: HENRY D. LAUSON

Council:
  ALEXANDER B. GUTMAN
  ROLLIN D. HOTCHKISS
  DAVID SHEMIN

1955–1956

President: COLIN M. MACLEOD
Vice-President: FRANK L. HORSFALL, JR.
Treasurer: CHANDLER McC. BROOKS
Secretary: RULON W. RAWSON

Council:
  ROLLIN D. HOTCHKISS
  DAVID SHEMIN
  ROBERT F. WATSON

1956–1957

President: FRANK L. HORSFALL, JR.
Vice-President: WILLIAM S. TILLETT
Treasurer: CHANDLER McC. BROOKS
Secretary: RULON W. RAWSON

Council:
  DAVID SHEMIN
  ROBERT F. WATSON
  ABRAHAM WHITE

1957–1958

President: WILLIAM S. TILLETT
Vice-President: ROLLIN D. HOTCHKISS
Treasurer: CHANDLER McC. BROOKS
Secretary: H. SHERWOOD LAWRENCE

Council:
  ROBERT F. WATSON
  ABRAHAM WHITE
  JOHN V. TAGGART

1958–1959

President: ROLLIN D. HOTCHKISS
Vice-President: ANDRE COURNAND
Treasurer: CHANDLER McC. BROOKS
Secretary: H. SHERWOOD LAWRENCE

Council:
  ABRAHAM WHITE
  JOHN V. TAGGART
  WALSH MCDERMOTT

1959–1960

President: ANDRE COURNAND
Vice-President: ROBERT F. PITTS
Treasurer: EDWARD J. HEHRE
Secretary: H. SHERWOOD LAWRENCE

Council:
  JOHN V. TAGGART
  WALSH MCDERMOTT
  ROBERT F. FURCHGOTT

1960–1961

President: ROBERT F. PITTS
Vice-President: DICKINSON W. RICHARDS
Treasurer: EDWARD J. HEHRE
Secretary: ALEXANDER G. BEARN

Council:
  WALSH MCDERMOTT
  ROBERT F. FURCHGOTT
  LUDWIG W. EICHNA

1961–1962

President: DICKINSON W. RICHARDS
Vice-President: PAUL WEISS
Treasurer: I. HERBERT SCHEINBERG
Secretary: ALEXANDER G. BEARN

Council:
ROBERT F. FURCHGOTT
LUDWIG W. EICHNA
EFRAIM RACKER

1962–1963

President: PAUL WEISS
Vice-President: ALEXANDER B. GUTMAN
Treasurer: I. HERBERT SCHEINBERG
Secretary: ALEXANDER G. BEARN

Council:
LUDWIG W. EICHNA
EFRAIM RACKER
ROGER L. GREIF

1963–1964

President: ALEXANDER B. GUTMAN
Vice-President: EDWARD L. TATUM
Treasurer: SAUL J. FARBER
Secretary: ALEXANDER G. BEARN

Council:
EFRAIM RACKER
ROGER L. GREIF
IRAVING M. LONDON

1964–1965

President: EDWARD TATUM
Vice-President: CHANDLER McC. BROOKS
Treasurer: SAUL J. FARBER
Secretary: RALPH L. ENGLE, JR.

Council:
ROGER L. GREIF
LEWIS THOMAS
IRVING M. LONDON

1965–1966

President: CHANDLER McC. BROOKS
Vice-President: ABRAHAM WHITE
Treasurer: SAUL J. FARBER
Secretary: RALPH L. ENGLE, JR.

Council:
IRVING M. LONDON
LEWIS THOMAS
GEORGE K. HIRST

1966–1967

President: ABRAHAM WHITE
Vice-President: RACHMIEL LEVINE
Treasurer: SAUL J. FARBER
Secretary: RALPH L. ENGLE, JR.

Council:
LEWIS THOMAS
GEORGE K. HIRST
DAVID NACHMANSOHN

1967–1968

President: RACHMIEL LEVINE
Vice-President: SAUL J. FARBER
Treasurer: PAUL A. MARKS
Secretary: RALPH L. ENGLE, JR.

Council:
GEORGE K. HIRST
DAVID NACHMANSOHN
MARTIN SONENBERG

1968–1969

President: SAUL J. FARBER
Vice-President: JOHN V. TAGGART
Treasurer: PAUL A. MARKS
Secretary: ELLIOTT F. OSSERMAN

Council:
DAVID NACHMANSOHN
MARTIN SONENBERG
HOWARD A. EDER

1969–1970

President: JOHN V. TAGGART
Vice-President: BERNARD L. HORECKER
Treasurer: PAUL A. MARKS
Secretary: ELLIOTT F. OSSERMAN

Council:
　MARTIN SONENBERG
　HOWARD A. EDER
　SAUL J. FARBER

1970–1971

President: BERNARD L. HORECKER
Vice-President: MACLYN MCCARTY
Treasurer: EDWARD C. FRANKLIN
Secretary: ELLIOTT F. OSSERMAN

Council:
　HOWARD A. EDER
　SAUL J. FARBER
　SOLOMON A. BERSON

1971–1972

President: MACLYN MCCARTY
Vice-President: ALEXANDER G. BEARN
Treasurer: EDWARD C. FRANKLIN
Secretary: ELLIOTT F. OSSERMAN

Council:
　SAUL J. FARBER
　SOLOMON A. BERSON
　HARRY EAGLE

1972–1973

President: ALEXANDER G. BEARN
Vice-President: PAUL A. MARKS
Treasurer: EDWARD C. FRANKLIN
Secretary: JOHN ZABRISKIE

Council:
　HARRY EAGLE
　JERARD HURWITZ

1973–1974

President: PAUL A. MARKS
Vice-President: IGOR TAMM
Treasurer: EDWARD C. FRANKLIN
Secretary: JOHN B. ZABRISKIE

Council:
　HARRY EAGLE
　CHARLOTTE FRIEND
　JERARD HURWITZ

1974–1975

President: IGOR TAMM
Vice-President: GERALD M. EDELMAN
Treasurer: STEPHEN I. MORSE
Secretary: JOHN B. ZABRISKIE

Council:
　JERARD HURWITZ
　H. SHERWOOD LAWRENCE
　CHARLOTTE FRIEND

1975–1976

President: GERALD M. EDELMAN
Vice-President: ELVIN A. KABAT
Treasurer: STEPHEN I. MORSE
Secretary: JOHN B. ZABRISKIE

Council:
　PAUL A. MARKS
　H. SHERWOOD LAWRENCE
　CHARLOTTE FRIEND

1976–1977

President: ELVIN A. KABAT
Vice-President: FRED PLUM
Treasurer: STEPHEN I. MORSE
Secretary: DONALD M. MARCUS

Council:
　H. SHERWOOD LAWRENCE
　PAUL A. MARKS
　BRUCE CUNNINGHAM

1977–1978

President: FRED PLUM	*Council:*
Vice-President: CHARLOTTE FRIEND	PAUL A. MARKS
Treasurer: STEPHEN I. MORSE	BRUCE CUNNINGHAM
Secretary: DONALD M. MARCUS	VITTORIO DEFENDI

1978–1979

President: CHARLOTTE FRIEND	*Council:*
Vice-President: MARTIN SONENBERG	BRUCE CUNNINGHAM
Treasurer: ALFRED STRACHER	VITTORIO DEFENDI
Secretary: DONALD M. MARCUS	DEWITT S. GOODMAN

1979–1980

President: MARTIN SONENBERG	*Council:*
Vice-President: KURT HIRSCHHORN	VITTORIO DEFENDI
Treasurer: ALFRED STRACHER	DEWITT S. GOODMAN
Secretary: EMIL C. GOTSCHLICH	ORA ROSEN

1980–1981

President: KURT HIRSCHHORN	*Council:*
Vice-President: GERALD WEISSMANN	RALPH NACHMAN
Treasurer: ALFRED STRACHER	DEWITT S. GOODMAN
Secretary: EMIL C. GOTSCHLICH	ORA ROSEN

1981–1982

President: GERALD WEISSMANN	*Council:*
Vice-President: DEWITT S. GOODMAN	KURT HIRSCHHORN
Treasurer: ALFRED STRACHER	RALPH L. NACHMAN
Secretary: EMIL C. GOTSCHLICH	ORA ROSEN

1982–1983

President: DEWITT S. GOODMAN	*Council:*
Vice-President: MATTHEW D. SCHARFF	KURT HIRSCHHORN
Treasurer: ALFRED STRACHER	RALPH L. NACHMAN
Secretary: EMIL C. GOTSCHLICH	GERALD WEISSMANN

1983–1984

President: MATTHEW D. SCHARFF	*Council:*
Vice-President: HAROLD S. GINSBERG	KURT HIRSCHHORN
Treasurer: RICHARD A. RIFKIND	GERALD WEISSMANN
Secretary: EMIL C. GOTSCHLICH	JAMES P. QUIGLEY

1984–1985

President: HAROLD S. GINSBERG	*Council:*
Vice-President: JAMES E. DARNELL	JAMES P. QUIGLEY
Treasurer: RICHARD A. RIFKIND	MATTHEW D. SCHARFF
Secretary: ROBERT J. DESNICK	GERALD WEISSMANN

1985–1986

President: JAMES E. DARNELL
Vice-President: DAVID SABATINI
Treasurer: RICHARD A. RIFKIND
Secretary: ROBERT J. DESNICK

Council:
JAMES P. QUIGLEY
MATTHEW D. SCHARFF
HAROLD S. GINSBERG

1986–1987

President: DAVID SABATINI
Vice-President: DONALD A. FISCHMAN
Treasurer: RICHARD A. RIFKIND
Secretary: ROBERT J. DESNICK

Council:
JAMES E. DARNELL
HAROLD S. GINSBERG
MATTHEW D. SCHARFF

1987–1988

President: DONALD A. FISCHMAN
Vice-President: JONATHAN R. WARNER
Treasurer: RICHARD A. RIFKIND
Secretary: ROBERT J. DESNICK

Council:
JAMES E. DARNELL
HAROLD S. GINSBERG
WILLIAM MCALLISTER
DAVID SABATINI

1988–1989

President: JONATHAN R. WARNER
Vice-President: ISIDORE S. EDELMAN
Treasurer: JOSEPH R. BERTINO
Secretary: ROBERT J. DESNICK

Council:
JAMES E. DARNELL
DONALD A. FISCHMAN
WILLIAM MCALLISTER
DAVID SABATINI

1989–1990

President: ISIDORE S. EDELMAN
Vice-President: DAVID J. LUCK
Treasurer: JOSEPH R. BERTINO
Secretary: PETER PALESE

Council:
DONALD A. FISCHMAN
ROCHELLE HIRSCHHORN
WILLIAM MCALLISTER
JONATHAN R. WARNER

1990–1991

President: DAVID J. LUCK
Vice-President: KENNETH BERNS
Treasurer: JOSEPH R. BERTINO
Secretary: PETER PALESE

Council:
ISIDORE S. EDELMAN
ROCHELLE HIRSCHHORN
WILLIAM MCALLISTER
JONATHAN R. WARNER

CUMULATIVE AUTHOR INDEX*

DR. STUART A. AARONSON, 1991–92 (h)
DR. JOHN J. ABEL, 1923–24 (d)
DR. JOHN ABELSON, 1989–90 (h)
PROF. J.D. ADAMI, 1906–07 (d)
DR. ROGER ADAMS, 1941–42 (d)
DR. THOMAS ADDIS, 1927–28 (d)
DR. JULIUS ADLER, 1976–77 (d)
DR. E.D. ADRIAN, 1931–32 (h)
DR. FULLER ALBRIGHT, 1942–43 (h)
DR. FRANZ ALEXANDER, 1930–31 (h)
DR. FREDERICK ALLEN, 1916–17 (d)
DR. JOHN F. ANDERSON, 1908–09 (d)
DR. R.J. ANDERSON, 1939–40 (d)
DR. CHRISTOPHER H. ANDREWS,
  1961–62 (h)
DR. CHRISTIAN B. ANFINSEN, 1965–66 (h)
PROF. G.V. ANREP, 1934–35 (h)
DR. CHARLES ARMSTRONG, 1940–41 (d)
DR. LUDWIG ASCHOFF, 1923–24 (d)
DR. LEON ASHER, 1922–23 (d)
DR. W.T. ASTBURY, 1950–51 (h)
DR. EDWIN ASTWOOD, 1944–45 (d)
DR. JOSEPH C. AUB, 1928–29 (d)
DR. K. FRANK AUSTEN, 1977–78 (h)
DR. RICHARD AXEL, 1983–84 (a)
DR. JULIUS AXELROD, 1971–72 (h)
DR. E.R. BALDWIN, 1914–15 (d)
DR. DAVID BALTIMORE, 1974–75 (h)
PROF. JOSEPH BARCROFT, 1921–22 (d)
DR. PHILIP BARD, 1921–22 (d)
DR. H.A. BARKER, 1949–50 (h)
DR. LEWELLYS BARKER, 1905–06 (d)
DR. JULIUS BAUER, 1932–33 (d)
PROF. WILLIAM M. BAYLISS, 1921–22 (d)
DR. DAVID BEACH, 1990–91 (h)
DR. FRANK BEACH, 1947–48 (h)
DR. GEORGE W. BEADLE, 1944–45 (h)
DR. ALEXANDER G. BEARN, 1974–75 (a)
DR. ALBERT BEHNKE, 1941–42 (h)
DR. BARUJ BENACERRAF, 1971–72 (a)
PROF. F.G. BENEDICT, 1906–07 (d)
DR. STANLEY BENEDICT, 1915–16 (d)
DR. STEPHEN J. BENKOVIC, 1991–92 (h)
DR. D. BENNETT, 1978–79 (a)

DR. M.V.L. BENNETT, 1982–83 (h)
PROF. R.R. BENSLEY, 1914–15 (d)
DR. SEYMOUR BENZER, 1960–61 (h)
Dr. PAUL BERG, 1971–72 (h)
DR. MAX BERGMANN, 1935–36 (d)
DR. SUNE BERGSTRÖM, 1974–75 (h)
DR. ROBERT W. BERLINER, 1958–59 (h)
DR. SOLOMAN A. BERSON, 1966–67 (d)
DR. MARCEL C. BESSIS, 1962–63 (h)
DR. C.H. BEST, 1940–41 (h)
DR. A. BEIDL, 1923–24 (d)
DR. RUPERT E. BILLINGHAM, 1966–67 (h)
DR. RICHARD J. BING, 1954–55 (a)
DR. J. MICHAEL BISHOP, 1982–83 (h)
DR. JOHN J. BITTNER, 1946–47 (d)
DR. ELIZABETH H. BLACKBURN,
  1990–91 (h)
PROF. FRANCIS G. BLAKE, 1934–35 (d)
DR. ALFRED BLALOCK, 1945–46 (d)
DR. GÜNTER BLOBEL, 1980–81 (a)
DR. KONRAD BLOCH, 1952–53 (h)
DR. BARRY R. BLOOM, 1988–89 (a)
DR. WALTER R. BLOOR, 1923–24 (d)
DR. DAVID BODIAN, 1956–57 (h)
DR. WALTER F. BODMER, 1976–77 (h)
DR. JAMES BONNER, 1952–53 (h)
DR. JULES BORDET, 1920–21 (d)
DR. DAVID BOTSTEIN, 1986–87 (h)
DR. WILLIAM T. BOVIE, 1922–23 (d)
DR. EDWARD A. BOYSE, 1971–72,
  1975–76 (h)
DR. STANLEY E. BRADLEY, 1959–60 (a)
DR. DANIEL BRANTON, 1981–82 (a)
DR. ARMIN C. BRAUN, 1960–61 (h)
DR. EUGENE BRAUNWALD, 1975–76 (h)
PROF. F. BREMER, (h)†
DR. RALPH L. BRINSTER, 1984–85 (h)
PROF. T.G. BRODIE, 1909–10 (d)
DR. DETLEV W. BRONK, 1933–34 (d)
DR. B. BROUWER, 1925–26 (d)
DR. DONALD D. BROWN, 1980–81 (a)
DR. MICHAEL S. BROWN, 1977–78 (h)
DR. WADE H. BROWN, 1928–29 (d)
DR. JOHN M. BUCHANAN, 1959–60 (h)

*(h), honorary; (a), active; (d), deceased.
†Did not present lecture because of World War II.

DR. R.E. DYER, 1933–34 (h)
DR. HARRY EAGLE, 1959–60 (a)
DR. E.M. EAST, 1930–31 (d)
DR. J.C. ECCLES, 1955–56 (h)
DR. GERALD M. EDELMAN, 1972–73 (a)
PROF. R.S. EDGAR, 1967–68 (h)
DR. DAVID L. EDSALL, 1907–08 (d)
DR. JOHN T. EDSALL, 1966–67 (h)
DR. WILLIAM EINTHOVEN, 1924–25 (d)
DR. HERMAN N. EISEN, 1964–65 (h)
DR. JOEL ELKES, 1961–62 (h)
DR. C.A. ELVEHJEM, 1939–40 (d)
DR. HAVEN EMERSON, 1954–55 (d)
DR. JOHN F. ENDERS, 1947–48, 1963–64 (h)
DR. BORIS EPHRUSSI, 1950–51 (h)
DR. JOSEPH ERLANGER, 1912–13, 1926–27 (h)
DR. EARL A. EVANS JR., 1943–44 (h)
DR. HERBERT M. EVANS, 1923–24 (h)
DR. JAMES EWING, 1907–08 (d)
DR. KNUD FABER, 1925–26 (d)
DR. W. FALTA, 1908–09 (d)
DR. W.O. FENN, 1927–28 (d)
DR. FRANK FENNER, 1956–57 (h)
DR. GERALD R. FINK, 1988–89 (a)
DR. H.O.L. FISCHER, 1944–45 (d)
DR. L.B. FLEXNER, 1951–52 (h)
DR. SIMON FLEXNER, 1911–12 (d)
DR. OTTO FOLIN, 1907–08, 1919–20 (d)
PROF. JOHN A. FORDYCE, 1914–15 (d)
DR. NELLIS B. FOSTER, 1920–21 (d)
DR. EDWARD FRANCIS, 1927–28 (d)
DR. THOMAS FRANCIS, JR., 1941–42 (d)
DR. H. FRAENKEL-CONRAT, 1956–57 (h)
DR. ROBERT T. FRANK, 1930–31 (d)
DR. EDWARD C. FRANKLIN, 1981–82 (d)
DR. DONALD S. FREDRICKSON, 1972–73 (h)
DR. IRWIN FRIDOVICH, 1983–84 (h)
DR. CARLOTTE FRIEND, 1976–77 (d)
DR. C. FROMAGEOT, 1953–56 (h)
DR. JOSEPH S. FRUTON, 1955–56 (a)
DR. JOHN F. FULTON, 1935–36 (d)
DR. E.J. FURSHPAN, 1980–81 (a)
DR. JACOB FURTH, 1967–68 (a)

DR. D. CARLETON GADJUSEK, 1976–77 (h)
DR. ERNEST F. GALE, 1955–56 (h)
DR. JOSEPH G. GALL, 1975–76 (h)
DR. T.F. GALLAGHER, 1956–57 (h)
DR. ROBERT C. GALLO, 1983–84 (h)
DR. JAMES L. GAMBLE, 1946–47 (d)
DR. HERBERT S. GASSER, 1936–37 (d)
DR. FREDERICK P. GAY, 1914–15, 1930–31 (d)
DR. WALTER J. GEHRING, 1985–86 (h)
DR. EUGENE M.K. GEILING, 1941–42 (d)
DR. ISIDORE GERSH, 1924–50 (h)
DR. GEORGE O. GEY, 1954–55 (d)
DR. JOHN H. GIBBON, 1957–58 (d)
DR. ALFRED G. GILMAN, 1989–90 (h)
DR. HARRY GOLDBLATT, 1937–38 (h)
DR. JOSEPH L. GOLDSTEIN, 1977–78 (h)
DR. ROBERT A. GOOD, 1971–72 (a)
DR. DEWITT S. GOODMAN, 1985–86 (a)
DR. EARNEST W. GOODPASTURE, 1929–30 (d)
DR. CARL W. GOTTSCHALK, 1962–63 (h)
DR. J. GOUGH, 1957–58 (h)
PROF. J.I. GOWANS, 1968–69 (h)
DR. EVARTS A. GRAHAM, 1923–24, 1933–34 (d)
DR. S. GRANICK, 1948–49 (h)
DR. DAVID E. GREEN, 1956–57 (h)
DR. HOWARD GREEN, 1978–79 (h)
DR. PAUL GREENGARD, 1979–80 (h)
PROF. R.A. GREGORY, 1968–69 (h)
DR. DONALD R. GRIFFIN, 1975–76 (h)
DR. JEROME GROSS, 1972–73 (h)
DR. ROGER GUILLEMIN, 1975–76 (h)
DR. I.C. GUNSALUS, 1949–50 (h)
DR. JOHN B. GURDON, 1973–74 (h)
DR. ALEXANDER B. GUTMAN, 1964–65 (a)
DR. J.S. HALDANE, 1916–17 (d)
DR. WILLIAM S. HALSTED, 1913–14 (d)
DR. H.J. HAMBURGER, 1922–23 (d)
DR. HIDESABURO HANAFUSA, 1979–80 (a)
DR. J.D. HARDY, 1953–54 (d)

PROF. W. MCKIM MARRIOTT, 1919–20 (d)
DR. E.K. MARSHALL, JR., 1929–30 (d)
DR. BRIAN W. MATTHEWS, 1986–86 (h)
DR. MANFRED M. MAYER, 1976–77 (h)
DR. DANIEL MAZIA, 1957–58 (h)
DR. MACLYN MCCARTY, 1969–70 (a)
PROF. E.V. MCCOLLUM, 1916–17 (d)
DR. WALSH MCDERMOTT, 1967–68 (h)
DR. HARDEN M. MCDONNELL, (h)
DR. W.D. MCELROY, 1955–56 (h)
DR. STEVEN LANIER MCKNIGHT, 1991–92 (h)
DR. PHILIP D. MCMASTER, 1941–42 (h)
DR. P.B. MEDAWAR, 1956–57 (h)
DR. WALTER J. MEEK, 1940–41 (d)
PROF. ALTON MEISTER, 1967–68 (h)
DR. S. J. MELTZER, 1906–07 (d)
PROF. LAFAYETTE B. MENDEL, 1905–06, 1914–15 (d)
DR. R. BRUCE MERRIFIELD, 1971–72 (h)
DR. HENRY METZGER, 1984–85 (h)
PROF. ADOLPH MEYER, 1909–10 (d)
PROF. HANS MEYER, 1905–06 (d)
DR. KARL MEYER, 1955–56 (h)
DR. K.F. MEYER, 1939–40 (d)
DR. OTTO MEYERHOF, 1922–23 (d)
DR. LEONOR MICHAELIS, 1926–27 (d)
DR. WILLIAM S. MILLER, 1924–25 (d)
PROF. CHARLES S. MINOT, 1905–06 (d)
DR. GEORGE R. MINOT, 1927–28 (d)
DR. BEATRICE MINTZ, 1975–76 (h)
DR. A.E. MIRSKY, 1950–51 (h)
DR. JACQUES MONOD, 1961–62 (h)
DR. CARL V. MOORE, 1958–59 (h)
DR. FRANCIS D. MOORE, 1956–57 (h)
DR. STANFORD MOORE, 1956–57 (h)
PROF. T.H. MORGAN, 1905–06 (d)
DR. GIUSEPPE MORUZZI, 1962–63 (h)
DR. J. HOWARD MUELLER, 1943–44 (d)
PROF. FRIEDRICH MULLER, 1906–07 (d)
DR. H.J. MULLER, 1947–48 (d)
DR. HANS MÜLLER-EBERHARD, 1970–71 (a)
PROF. JOHN R. MURLIN, 1916–17 (d)
DR. W.P. MURPHY, 1927–28 (d)
DR. DAVID NACHMANSOHN, 1953–54 (h)

DR. F.R. NAGER, 1925–26 (d)
DR. DANIEL NATHANS, 1974–75 (h)
DR. JAMES V. NEEL, 1960–61 (h)
DR. ELIZABETH F. NEUFELD, 1979–80 (h)
DR. FRED NEUFELD, 1926–27 (d)
SIR ARTHUR NEWSHOLME, 1920–21 (d)
DR. MARSHALL W. NIRENBERG, 1963–64 (h)
DR. HIDEYO NOGUVHI, 1915–16 (d)
DR. HARRY F. NOLLER, 1988–89 (a)
DR. JOHN H. NORTHROP, 1925–26, 1934–35 (d)
DR. G.J.V. NOSSAL, 1967–68 (h)
PROF. FREDERICK G. NOVY, 1934–35 (d)
DR. SHOSAKU NUMA, 1987–88 (a)
DR. RUTH S. NUSSENZWEIG, 1982–83 (a)
DR. VICTOR NUSSENZWEIG, 1982–83 (a)
DR. CHRISTIANE NÜSSLEIN-VOLHARD, 1990–91 (h)
PROF. GEORGE H.F. NUTTALL, 1912–13 (d)
DR. SEVERO OCHOA, 1950–51 (a)
DR. LLOYD J. OLD, 1971–72, 1975–76 (h)
DR. JOHN OLIPHANT, 1943–44 (d)
DR. JEAN OLIVER, 1944–45 (h)
DR. BERT W. O'MALLEY, 1976–77 (h)
DR. J.L. ONCLEY, 1954–55 (h)
DR. EUGENE L. OPIE, 1909–10, 1928–29, 1954–55 (d)
DR. STUART H. ORKIN, 1987–88 (a)
PROF. HENRY F. OSBORN, 1911–12 (d)
DR. MARY JANE OSBORN, 1982–83 (h)
DR. THOMAS B. OSBORNE, 1910–11 (d)
DR. WINTHROP J.V. OSTERHOUT, 1921–22, 1929–30 (d)
DR. GEORGE E. PALADE, 1961–62 (a)
DR. A.M. PAPPENHEIMER, JR., 1956–57, 1980–81 (a)
DR. JOHN R. PAPPENHEIMER, 1965–66 (a)
PROF. ARTHUR B. PARDEE, 1969–70 (h)
DR. EDWARDS A. PARK, 1938–39 (d)

Prof. W.H. Park, 1905–06 (d)
Prof. G.H. Parker, 1913–14 (d)
Dr. Stewart Paton, 1917–19 (d)
Dr. John R. Paul, 1942–43 (d)
Dr. L. Pauling, 1953–54 (h)
Dr. Francis W. Peabody, 1916–17 (d)
Prof. Richard M. Pearce, 1909–10 (d)
Dr. Raymond Pearl, 1921–22 (d)
Dr. William Stanley Peart, 1977–78 (h)
Dr. Wilder Penfield, 1936–37 (d)
Dr. M.F. Perutz, 1967–68 (h)
Dr. John P. Peters, 1937–38 (d)
Dr. W.H. Peterson, 1946–47 (d)
Dr. David C. Phillips, 1970–71 (h)
Dr. Ernst P. Pick, 1929–30 (h)
Dr. Ludwig Pick, 1931–32 (d)
Dr. Gregory Pincus, 1966–67 (d)
Dr. Clemens Pirquet, 1921–22 (d)
Dr. Colin Pitendrigh, 1960–61 (h)
Dr. Robert Pitts, 1952–53 (d)
Dr. A. Policard, 1931–32 (h)
Prof. George J. Popjak, 1969–70 (h)
Dr. Keith R. Porter, 1955–56 (a)
Prof. Rodney R. Porter, 1969–70 (h)
Dr. W.T. Porter, 1906–07, 1917–19 (d)
Dr. Stanley B. Prusiner, 1991–92 (h)
Dr. Mark Ptashne, 1973–74 (h)
Dr. T. T. Puck, 1958–59 (h)
Dr. J. J. Putnam, 1911–12 (d)
Dr. Vincent R. Racaniello, 1991–92 (h)
Dr. Efraim Racker, 1955–56 (a)
Dr. Hermann Rahn, 1958–59 (h)
Dr. Charles H. Rammelkamp, Jr., 1955–56 (h)
Dr. S. Walter Ranson, 1937–37 (d)
Dr. Kenneth B. Raper, 1961–62 (h)
Dr. Alexander Rich, 1982–83 (a)
Dr. Arnold R. Rich, 1946–57 (d)
Prof. Alfred N. Richards, 1920–21, 1934–35 (d)
Dr. Dickinson W. Richards, 1943–44 (h)
Prof. Theodore W. Richards, 1911–12 (d)
Dr. Curt P. Richter, 1942–43 (h)

Dr. D. Rittenberg, 1948–49 (d)
Dr. Thomas M. Rivers, 1933–34 (d)
Dr. William Robbins, 1942–43 (h)
Dr. O.H. Robertson, 1942–43 (d)
Prof. William C. Rose, 1934–35 (h)
Prof. Ora Mendelsohn Rosen, 1986–87 (a)
Dr. M.J. Rosenau, 1908–09 (d)
Dr. Russell Ross, 1981–82 (a)
Dr. Michael G. Rossmann, 1987–88 (a)
Dr. Jesse Roth, 1981–82 (a)
Dr. James E. Rothman, 1990–91 (h)
Dr. F.J.W. Roughton, 1943–44 (h)
Dr. Peyton Rous, 1935–36 (d)
Dr. Wallace P. Rowe, 1975–76 (h)
Dr. Gerald M. Rubin, 1987–88 (a)
Dr. Harry Rubin, 1965–66 (h)
Prof. Max Rubner, 1912–13 (d)
Dr. Frank H. Ruddle, 1973–74 (h)
Dr. John Runnstrom, 1950–51 (h)
Dr. Errki Ruoslahti, 1988–89 (a)
Major Frederick F. Russell, 1912–13 (d)
Dr. F.R. Sabin, 1915–16 (d)
Dr. Leo Sachs, 1972–73 (h)
Dr. Ruth Sager, 1982–83 (h)
Dr. Bengt Samuelsson, 1979–80 (h)
Dr. Wilbur A. Sawyer, 1934–35 (d)
Dr. Howard Schachman, 1972–73 (h)
Prof. E.A. Schafer, 1907–08 (d)
Dr. Gottfried Schatz, 1989–90 (h)
Dr. Robert T. Schimke, 1980–81 (h)
Dr. Matthew D. Scharff, 1973–74 (a)
Dr. Harold W. Scheraga, 1967–68 (h)
Dr. Bela Schick, 1922–23 (h)
Dr. Robert Schleif, 1988–89 (a)
Dr. Oscar Schloss, 1924–25 (d)
Dr. Stuart F. Schlossman, 1983–84 (h)
Prof. Adolph Schmidt, 1913–14 (d)
Dr. Carl F. Schmidt, 1948–49 (h)
Dr. Knut Schmidt-Neilsen, 1962–63 (h)

ACTIVE MEMBERS

DR. BENT AASTED
DR. ALAN ADEREM
DR. EWARD H. AHRENS*
DR. PHILIP AISEN
DR. SALAH AL-ASKARI
DR. QAIS AL-AWQATI
DR. ANTHONY A. ALBANESE*
DR. EMMA G. ALLEN
DR. NORMAN R. ALPERT
DR. BLANCHE PEARL ALTER
DR. NORMAN ALTSZULER
DR. BURTON M. ALTURA
DR. HELEN ANDERSON
DR. GIUSEPPE A. ANDRES
DR. REGINALD M. ARCHIBALD*
DR. HIROSHI ASANUMA
DR. AMIR ASKARI
DR. ARLEEN D. AUERBACH
DR. ARTHUR H. AUFSES, JR.
DR. PETER A.M. AULD
DR. ROBERT AUSTRIAN*
DR. THEODORE W. AVRUSKIN
DR. D. ROBERT AXELROD*
DR. STEPHEN M. AYRES
DR. EFRAIN CHARLES AZMITIA
DR. ROSTOM BABLANIAN
DR. RICHARD A. BADER
DR. DAVID S. BALDWIN
DR. AMIYA BANERJEE
DR. SACHCHIDANANDA BANERJEE*
DR. NILS U. BANG
DR. ARTHUR BANK
DR. S.B. BARKER*
DR. JAMES J. BARONDESS
DR. JEREMIAH A. BARONDESS
DR. BRUCE A. BARRON
DR. JACK R. BATTISTO*
DR. HAGAN BAYLEY
DR. ALEXANDER G. BEARN*
DR. A. ROBERT BECK
DR. CARL G. BECKER
DR. JOSEPH W. BECKER
DR. BERTRAND M. BELL

DR. BARUJ BENACERRAF
DR. BERNARD BENJAMIN*
DR. BRY BENJAMIN
DR. THOMAS P. BENNETT
DR. GORDON D. BENSON
DR. RICHARD BERESFORD
DR. LAWRENCE BERGER
DR. GREGORY I. BERK
DR. PAUL D. BERK
DR. JAMES I. BERKMAN*
DR. BRIAN BERMAN
DR. ALAN W. BERNHEIMER*
DR. HARRIET P. BERNHEIMER*
DR. KENNETH I. BERNS
DR. L.H. BERNSTEIN
DR. CARL A. BERNTSEN
DR. JOSEPH R. BERTINO
DR. JOHN F. BERTLES
DR. R.J. BING*
DR. DAVID BISHOP
DR. DAVID H. BLANKENHORN*
DR. RICHARD W. BLIDE
DR. ANDREW BLITZER
DR. GUNTER K.J. BLOBEL
DR. KONRAD BLOCH*
DR. BERNARD H. BOAL
DR. RICHARD STEVEN BOCKMAN
DR. DIETHELM H. BOEHME
DR. ALFRED J. BOLLET
DR. RICHARD J. BONFORTE
DR. ADELE L. BOSKEY
DR. BARBARA H. BOWMAN
DR. ROBERT J. BOYLAN
DR. RICHARD C. BOZIAN*
DR. THOMAS B. BRADLEY, JR.
DR. LEON BRADLOW
DR. JO ANNE BRASEL
DR. GOODWIN M. BREININ
DR. ESTHER BRESLOW
DR. JAN L. BRESLOW
DR. ROBIN W. BRIEHL
DR. STANLEY A. BRILLER*
DR. ANNE M. BRISCOE*

*Life Member.

Dr. Felix Bronner
Dr. Mark S. Brower
Dr. Steven T. Brower
Dr. Audrey K. Brown*
Dr. John L. Brown*
Dr. Howard C. Bruenn*
Dr. Elmer Brummer
Dr. Thomas M. Buchanan*
Dr. Doris J. Bucher
Dr. Joseph A. Buda*
Dr. Richard Burger
Dr. Roger M. Burnett
Dr. John J. Burns
Dr. Vincent P. Butler, Jr.
Dr. Joel N. Buxbaum
Dr. Peter R.B. Caldwell
Dr. Robert Cancro
Dr. Robert E. Canfield
Dr. Paul J. Cannon
Dr. Peter W. Carmel
Dr. Hugh J. Carroll
Dr. Anne C. Carter*
Dr. David M. Carter
Dr. J. Casals-Ariet
Dr. David B. Case
Dr. Joan I. Casey
Dr. Peter Cervoni
Dr. Raju S.K. Chaganti
Dr. Robert Warner Chambers*
Dr. W.Y. Chan
Dr. Joseph P. Chandler*
Dr. Moses Victor Chao
Dr. Merrill W. Chase*
Dr. Norman E. Chase
Dr. Herbert Chasis*
Dr. Leonard Chess
Dr. David Chi
Dr. Raul Chiesa
Dr. Francis P. Chinard*
Dr. Yong Sung Choi
Dr. Purnell W. Choppin
Dr. Charles L. Christian
Dr. Judith K. Christman

Dr. Jacob Churg*
Dr. Dennis J. Cleri
Dr. Hartwig Cleve
Dr. Leighton E. Cluff*
Dr. Gerald Cohen
Dr. Ira S. Cohen
Dr. Mildred Cohn*
Dr. Zanvil A. Cohn*
Dr. Randolph P. Cole
Dr. Morton Coleman
Dr. David R. Colman
Dr. Lawrence A. Cone
Dr. Stephen Connolly
Dr. Jean L. Cook*
Dr. Stuart D. Cook
Dr. Norman S. Cooper*
Dr. Armand F. Cortese
Dr. Andre Cournand*
Dr. David Cowen*
Dr. John P. Craig
Dr. Bruce N. Cronstein
Dr. Richard J. Cross*
Dr. Kathryn L. Crossin
Dr. Bruce Cunningham
Dr. Dorothy J. Cunningham
Dr. Tom Curran
Dr. Richard F. D'Amato
Dr. James E. Darnell
Dr. John R. David
Dr. Jean Davignon
Dr. Robert P. Davis
Dr. Paul F. De Gara*
Dr. Thomas J. Degnan
Dr. Ralph B. Dell
Dr. Robert J. Dellenback
Dr. John R. Denton
Dr. Robert J. Desnick
Dr. Dickson D. Despommier
Dr. Bhanumas Dharmgrongartama
Dr. Alberto Di Donato
Dr. Joseph R. Di Palma*
Dr. Alexandra S. Dimich
Dr. Ann M. Dnistrian

*Life Member.

Dr. Alvin M. Donnenfeld
Dr. Philip J. Dorman
Dr. Gordon W. Douglas
Dr. R. Gordon Douglas, Jr.
Dr. Steven D. Douglas
Dr. Peter Clowes Dowling
Dr. Arnold Drapkin
Dr. Paul Dreizen
Dr. David T. Dresdale*
Dr. M. Catherine Driscoll
Dr. Lewis M. Drusin
Dr. Ronald E. Drusin
Dr. A.E. Dumont
Dr. Bo Dupont
Dr. Murray Dworetsky*
Dr. Harry Eagle*
Dr. Gerald M. Edelman
Dr. Isidore S. Edelman
Dr. Norman Edelman
Dr. Richard L. Edelson
Dr. Mark L. Edwards
Dr. Hans J. Eggers*
Dr. Kathryn H. Ehlers
Dr. Ludwig W. Eichna*
Dr. Robert P. Eisinger
Dr. John T. Ellis
Dr. Peter Elsbach
Dr. Samuel Elster
Dr. Mary Allen Engle*
Dr. Ralph L. Engle, Jr.*
Dr. Bernard F. Erlanger
Dr. Diane Esposito
Dr. Solomon Estren
Dr. Earl A. Evans, Jr.*
Dr. Ronald B. Faanes
Dr. Saul J. Farber*
Dr. Mehdi Farhangi
Dr. Lee E. Farr*
Dr. Daniel S. Feldman*
Dr. Bernard N. Fields*
Dr. Laurence Finberg
Dr. Arthur D. Finck
Dr. Donald Fischman

Dr. Arthur Fishberg*
Dr. Paul B. Fisher
Dr. Saul H. Fisher
Dr. Patrick Fitzgerald
Dr. Raul Fleischmajer
Dr. Howard B. Fleit
Dr. Kathleen M. Foley
Dr. Arthur C. Fox
Dr. Blas Frangione
Dr. Carl Frasch
Dr. Blair A. Fraser
Dr. Irwin M. Freedberg
Dr. Aaron D. Freedman*
Dr. Michael L. Freedman
Dr. Alvin Freiman
Dr. Arnold J. Friedhoff
Dr. Ralph Friedlander*
Dr. Eli A. Friedman
Dr. Stanley Friedman
Dr. Wilhelm R. Frisell
Dr. Steven Fruchtman
Dr. Joseph S. Fruton*
Dr. Hiroyoshi Fujita
Dr. Mulford Fulop
Dr. Robert F. Furchgott*
Dr. Mark E. Furth
Dr. Palmer F. Futcher*
Dr. Jacques L. Gabrilove*
Dr. Morton Galdston*
Dr. W. Einar Gall
Dr. Samuel Gandy
Dr. Martin Gardy
Dr. Frederick Gates III
Dr. Mario Gaudino*
Dr. Lester M. Geller
Dr. Donald A. Gerber
Dr. James German III
Dr. Marvin Gershengorn
Dr. Edward L. Gershey
Dr. Michael D. Gershon
Dr. Elizabeth C. Gerst
Dr. Menard M. Gertler*
Dr. Allan Gibofsky

*Life Member.

Dr. Irma Gigli
Dr. Harriet S. Gilbert
Dr. Helena Gilder
Dr. Charles Gilvarg
Dr. James Z. Ginos
Dr. Harold S. Ginsberg*
Dr. Henry Ginsberg
Dr. Sheldon Glabman
Dr. Warren Glaser
Dr. Leonard Glass
Dr. Ephraim Glassman*
Dr. Vincent V. Glaviano*
Dr. Robert Morris Glickman
Dr. David L. Globus
Dr. Martin J. Glynn*
Dr. Gabriel C. Godman*
Dr. C. Nigel Godson
Dr. Edmond A. Goidl
Dr. Allen M. Gold
Dr. Jonathan W.M. Gold
Dr. Leslie I. Gold
Dr. Allan R. Goldberg
Dr. Ira J. Goldberg
Dr. Lewis Goldfrank
Dr. Roberta Goldring
Dr. Laura T. Goldsmith
Dr. N. Goldsmith*
Dr. David A. Goldstein
Dr. Gideon Goldstein
Dr. Marvin H. Goldstein
Dr. Robert Goldstein
Dr. Robert A. Good*
Dr. Alvin J. Gordon*
Dr. Harry H. Gordon*
Dr. Richard Gorlin
Dr. Emil C. Gotschlich
Dr. Eugene L. Gottfried
Dr. Otto Gotze
Dr. Dicran Goulian
Dr. R.F. Grady
Dr. Lester Grant
Dr. Jack P. Green
Dr. Lowell M. Greenbaum

Dr. Jeffrey B. Greene
Dr. Olga Greengard
Dr. Ezra M. Greenspan
Dr. John D. Gregory*
Dr. Roger L. Greif*
Dr. Anthony J. Grieco
Dr. Joel Grinker
Dr. Arthur Grishman
Dr. David Grob*
Dr. Melvin M. Grumbach
Dr. Joseph J. Guarneri
Dr. Peter M. Guida
Dr. Guido Guidotti
Dr. Subhash Gulati
Dr. Sidney Gutstein
Dr. Gail S. Habicht
Dr. David V. Habif*
Dr. Susan Hadley*
Dr. Jack W.C. Hagstrom
Dr. Kathleen A. Haines
Dr. David P. Hajjar
Dr. Katherine A. Hajjar
Dr. James B. Hamilton*
Dr. Leonard D. Hamilton
Dr. Hideasburo Hanafusa
Dr. Eugene S. Handler
Dr. Ken Harewood
Dr. Peter C. Harpel
Dr. George A. Hashim
Dr. Sami Hashim
Dr. Victor Hatcher
Dr. A. Daniel Hauser
Dr. Richard Hawkins
Dr. Richard M. Hays
Dr. Michael Heidelberger*
Dr. Samuel Hellman
Dr. Wylie C. Hembree
Dr. Michael V. Herman
Dr. Margaret W. Hilgartner
Dr. C.H.W. Hirs
Dr. Jacob I. Hirsch
Dr. James G. Hirsch*
Dr. Jules Hirsch

*Life Member.

DR. KURT HIRSCHHORN
DR. ROCHELLE HIRSCHHORN
DR. GEORGE K. HIRST*
DR. PAUL HOCHSTEIN
DR. DAVID S. HODES
DR. RAYMOND F. HOLDEN, JR.*
DR. PETER R. HOLT
DR. ROBERT S. HOLZMAN
DR. WILLIAM H. HORNER*
DR. MARSHALL S. HORWITZ
DR. ROLLIN D. HOTCHKISS*
DR. S. STEVEN HOTTA
DR. LING-LING HSIEH
DR. HOWARD H.T. HSU
DR. KONRAD C. HSU
DR. W.N. HUBBARD, JR.
DR. JERARD HURWITZ
DR. DORRIS J. HUTCHINSON*
DR. THOMAS HUTTEROTH
DR. JULIANNE IMPERATO-MCGINLEY
DR. LAURA INSELMAN
DR. HARRY L. IOACHIM
DR. NORMAN J. ISAACS
DR. HENRY D. ISENBERG
DR. FARAMARZ ISMAIL-BEIGI
DR. FUYUKI IWASA
DR. RICHARD W. JACKSON*
DR. JERRY C. JACOBS
DR. ERIC A. JAFFE
DR. ERNST R. JAFFE*
DR. HERBERT JAFFE*
DR. ALFONSO H. JANOSKI
DR. HENRY D. JANOWITZ*
DR. SAUL JARCHO*
DR. CHARLES I. JAROWSKI
DR. JAMSHID JAVID
DR. NORMAN B. JAVITT
DR. S. MICHAL JAZWINSKI
DR. GRAHAM H. JEFFRIES
DR. ALAN JOHNSON*
DR. WARREN D. JOHNSON, JR.
DR. ALAN S. JOSEPHSON
DR. RICHARD JOVE

DR. DAVID B. KABACK
DR. RONALD KABACK
DR. ELVIN A. KABAT*
DR. LAWRENCE J. KAGAN
DR. MARTIN L. KAHN
DR. MELVIN KAHN
DR. THOMAS KAHN
DR. MIKIO KAMIYAMA
DR. SANDRA KAMMERMAN
DR. ERIC R. KANDEL
DR. BARRY H. KAPLAN
DR. GILLA KAPLAN
DR. KAREN L. KAPLAN
DR. ATTALLAH KAPPAS
DR. ARTHUR KARANAS
DR. ARTHUR KARLIN
DR. STUART S. KASSAN
DR. MICHAEL KATZ
DR. MITCHELL A. KATZ
DR. HANS KAUNITZ
DR. HERBERT J. KAYDEN
DR. DONALD KAYE
DR. GORDON I. KAYE
DR. AARON KELLNER*
DR. LEO KESNER
DR. GERALD T. KEUSCH
DR. EDWIN D. KILBOURNE*
DR. YOON BERM KIM
DR. ROBERT P. KIMBERLEY
DR. TOM KINDT
DR. DONALD WEST KING
DR. GLENN C. KING*
DR. DAVID W. KINNE
DR. DAVID G. KLAPPER
DR. JANIS V. KLAVINS
DR. BERNARD KLEIN*
DR. HERBERT A. KLEIN
DR. ALBRECHT K. KLEINSCHMIDT
DR. PERCY KLINGENSTEIN*
DR. JOSEPH A. KOCHEN
DR. SAMUEL S. KOIDE
DR. KIYOMI KOIZUMI
DR. LEVY KOPELOVICH

*Life Member.

Dr. Arthur Kornberg*
Dr. Irvin M. Korr*
Dr. Charles E. Kossmann*
Dr. Ione A. Kourides
Dr. O. Dhodanand Kowlessar
Dr. Thomas R. Kozel
Dr. Philip J. Kozinn
Dr. Irwin H. Krakoff
Dr. Lawrence Krakoff
Dr. Richard M. Krause
Dr. Alfred N. Krauss
Dr. Richard Kravath
Dr. Gert Kreibich
Dr. Norman Kretchmer*
Dr. Stephen Krop*
Dr. Saul Krugman*
Dr. Raju S. Kucherlapati
Dr. Friedrich Kueppers
Dr. Sherman Kupfer
Dr. H.S. Kupperman
Dr. Marvin Kuschner
Dr. Sau-Ping Kwan
Dr. Elizabeth H. Lacy
Dr. Robert G. Lahita
Dr. Chun-Yen Lai
Dr. Michael E. Lamm
Dr. Frank R. Landsberger
Dr. William B. Langan*
Dr. Philip Lanzkowsky
Dr. Etienne Y. Lasfargues
Dr. John Lattimer*
Dr. Leroy S. Lavine
Dr. H. Sherwood Lawrence*
Dr. Richard W. Lawton*
Dr. Paul B. Lazarow
Dr. Robert A. Lazzarini
Dr. Stanley L. Lee*
Dr. Sylvia Lee-Huang
Dr. Albert M. Lefkovits
Dr. Thomas J.A. Lehman
Dr. David Lehr*
Dr. Gerard M. Lehrer
Dr. Edgar Leifer

Dr. Roger L. Lerner
Dr. E. Carwile Leroy
Dr. Stephen H. Leslie*
Dr. Gerson J. Lesnick*
Dr. Harry H. Le Veen*
Dr. Richard D. Levere
Dr. Harold A. Levey
Dr. Roberto Levi
Dr. Richard I. Levin
Dr. Robert A. Levine
Dr. Marvin F. Levitt*
Dr. Lester M. Levy
Dr. Marjorie Lewisohn*
Dr. Charles S. Lieber
Dr. Kenneth V. Lieberman
Dr. Seymour Lieberman*
Dr. Martin R. Liebowitz
Dr. Frank Lilly
Dr. Stanley L. Lin
Dr. Alfred S.C. Ling
Dr. George Lipkin
Dr. Martin Lipkin
Dr. Arthur H. Livermore*
Dr. Rodolfo Llinas
Dr. John N. Loeb
Dr. Werner R. Loewenstein
Dr. Irving M. London*
Dr. Morris London*
Dr. Barbara W. Low
Dr. Jerome Lowenstein
Dr. Oliver Lowry*
Dr. Daniel S. Lukas*
Dr. Carol J. Lusty
Dr. Daniel Macken
Dr. Thomas Magill*
Dr. Richard J. Mahler
Dr. Jacob V. Maizel, Jr.
Dr. William Manger
Dr. Belur N. Manjula
Dr. Mart Mannik
Dr. James M. Manning
Dr. Karl Maramorosch*
Dr. Aaron J. Marcus

*Life Member.

DR. PHILIP I. MARCUS
DR. CATHERINE T. MARINO
DR. MORTON MARKS*
DR. PAUL A. MARKS
DR. DANIEL S. MARTIN
DR. BENTO MASCARENHAS
DR. EDMUND B. MASUROVSKY
DR. LEONARD M. MATTES
DR. ROBERT MATZ
DR. PAUL MAURER*
DR. MORTON MAXWELL*
DR. KLAUS MAYER
DR. VALENTINO D.B. MAZZIA
DR. KENNETH S. MCCARTY
DR. MACLYN MCCARTY*
DR. JOHN CHARLES MCGIFF
DR. PAUL R. MCHUGH
DR. JAMES J. MCSHARRY
DR. ROBERT MCVIE
DR. JOHN G. MEARS
DR. EDWARD MEILMAN
DR. GILBERT W. MELLIN
DR. ROBERT B. MELLINS
DR. VINCENT J. MERLUZZI
DR. JOSEPH MICHL
DR. DONNA MILDVAN
DR. FREDERICK MILLER
DR. MYRON MILLER
DR. PETER MODEL
DR. RICHARD C. MOHS
DR. CARL MONDER
DR. MALCOLM A.S. MOORE
DR. TAKASHI MORIMOTO
DR. AKIRA MORISHIMA
DR. THOMAS MORRIS
DR. JOHN MORRISON
DR. KEVIN P. MORRISSEY
DR. MELVIN L. MOSS
DR. HARRY MOST*
DR. RICHARD W. MOYER
DR. GILBERT H. MUDGE*
DR. HANS J. MUELLER-EBERHARD
DR. URSULA MÜLLER-EBERHARD

DR. M. LOIS MURPHY
DR. BEN ADDISON MURRAY
DR. W.P. LAIRD MYERS*
DR. RALPH L. NACHMAN
DR. HARRIS M. NAGLER
DR. TATSUJI NAMBA
DR. MORTON NATHANSON*
DR. GERALD NATHENSON
DR. STANLEY G. NATHENSON
DR. BRIAN A. NAUGHTON
DR. JAMES V. NEEL
DR. AARON R. NEVIN
DR. MARIA NEW
DR. CAROL SHAW NEWLON
DR. WARREN W. NICHOLS
DR. JOHN F. NICHOLSON
DR. JULIAN NIEMETZ
DR. ELENA O. NIGHTINGALE
DR. JEROME S. NISSELBAUM
DR. CHARLES NOBACK*
DR. ROBERT NOLAN
DR. G.J.V. NOSSAL
DR. NEIL J. NUSBAUM
DR. VICTOR NUSSENZWEIG
DR. MANUEL OCHOA, JR.*
DR. SEVERO OCHOA*
DR. HERBERT F. OETTGEN
DR. ALLEN I. OLIFF
DR. INGRITH J. DEYRUP OLSEN*
DR. CARL A. OLSSON
DR. PETER D. ORAHOVATS*
DR. IRWIN ORESKES
DR. ERNEST V. ORSI*
DR. ZOLTAN OVARY
DR. NORBERT I.A. OVERWEG
DR. JOHN OWEN
DR. HARVEY L. OZER
DR. ELIZABETH PAIETTA
DR. GEORGE E. PALADE*
DR. PETER PALESE
DR. KRZYSZTOF PANKIEWKZ
DR. GEORGE D. PAPPAS
DR. ALVIN M. PAPPENHEIMER, JR.

*Life Member.

Dr. John R. Pappenheimer*
Dr. Jean Papps*
Dr. Constance M. Park
Dr. Pedro Pasik
Dr. Mark W. Pasmantier
Dr. Gavril W. Pasternak
Dr. Philip Y. Paterson*
Dr. Jaygonda R. Patil
Dr. Pierluigi Patriarca
Dr. Christian C. Patrick
Dr. Mary Ann Payne*
Dr. William Stanley Peart
Dr. Elinor I.B. Peerschke
Dr. Robert G. Pergolizzi
Dr. Demetrius Pertsemlidis
Dr. Sidney Pestka
Dr. Karl H. Pfenninger
Dr. Lennart Philipson*
Dr. Julia M. Phillips-Quagliata
Dr. John G. Pierce*
Dr. Matthew Pincus
Dr. Kermit L. Pines*
Dr. Xavier Pi-Sunyer
Dr. Walter F. Pizzi
Dr. Beatriz G.T. Pogo
Dr. Kenneth B. Pomerantz
Dr. Edwin A. Popenoe*
Dr. Keith R. Porter*
Dr. Jerome G. Porush
Dr. Jerome B. Posner
Dr. Marshall P. Primack
Dr. John F. Prudden
Dr. Lawrence Prutkin
Dr. Ellen Pure
Dr. Dominick P. Purpura
Dr. Franco Quagliata
Dr. Enrique M. Rabellino
Dr. Efraim Racker*
Dr. Shalom Rackovsky
Dr. Bertha Rader
Dr. Henry T. Randall*
Dr. Helen M. Ranney*
Dr. Felix T. Rapaport

Dr. Aaron Rausen
Dr. Jeffrey V. Ravetch
Dr. Lawrence W. Raymond
Dr. George G. Reader*
Dr. Walter Redisch*
Dr. Colvin M. Redman
Dr. George E. Reed
Dr. George N. Reeke, Jr.
Dr. Gabrielle H. Reem
Dr. Westley Reeves
Dr. Joan Reibman
Dr. Marcus M. Reidenberg
Dr. Joseph Reilly*
Dr. Leopold Reiner*
Dr. Donald J. Reis
Dr. G.W. Richter*
Dr. Ronald F. Rieder
Dr. Richard A. Rifkind
Dr. Robert R. Riggio
Dr. Walter F. Riker*
Dr. David A. Ringle*
Dr. Marcos Rivelis
Dr. Richard S. Rivlin
Dr. Jay Roberts
Dr. Alan G. Robinson
Dr. Robert G. Roeder
Dr. Murray D. Rosenberg
Dr. William Rosner
Dr. James Rothman
Dr. M.A. Rothschild*
Dr. Lewis P. Rowland
Dr. Paul C. Royce
Dr. B.A. Rubin*
Dr. Daniel Rudman
Dr. David Sabatini
Dr. David B. Sachar
Dr. Ruth Sager
Dr. Gerald Salen
Dr. Letty G.M. Salentijn
Dr. Jane E. Salmon
Dr. Milton R.J. Salton
Dr. Paul Samuel
Dr. Stanley Samuels
Dr. John I. Sandson

*Life Member.

Dr. Nanette F. Santoro
Dr. Shigeru Sassa
Dr. Arthur Sawitsky
Dr. Russell W. Schaedler
Dr. Matthew D. Scharff
Dr. Barbara M. Scher
Dr. William Scher
Dr. Donald J. Scherl
Dr. Lawrence Scherr
Dr. Peter B. Schiff
Dr. Gerald Schiffman
Dr. Edward B. Schlesinger*
Dr. Donald H. Schmidt
Dr. Howard A. Schneider*
Dr. Vern L. Schramm
Dr. Rise Schwab
Dr. Ernest Schwartz
Dr. Irving L. Schwartz*
Dr. James H. Schwartz
Dr. Morton K. Schwartz
Dr. Richard H. Schwartz
Dr. David Schwimmer
Dr. John J. Sciarra
Dr. James J. Sciubba
Dr. William A. Scott
Dr. J.C. Scott-Baker*
Dr. Sheldon J. Segal
Dr. Pravinkumar B. Sehgal
Dr. Irving Seidman
Dr. Samuel Seifter*
Dr. Ewald Selkurt*
Dr. Fabio Sereni*
Dr. Robert E. Shank
Dr. Warren B. Shapiro
Dr. Michael Shelanski
Dr. Raymond L. Sherman
Dr. Joyce E. Shriver
Dr. Bernard J. Sicuranza
Dr. M.A.Q. Siddiqui
Dr. Edward J. Siden
Dr. George J. Siegel
Dr. Morris Siegel*
Dr. Richard T. Silver

Dr. Martin E. Silverstein
Dr. Roy L. Silverstein
Dr. Saul J. Silverstein
Dr. Michael S. Simberkoff
Dr. Eric J. Simon
Dr. Joe L. Simpson
Dr. Anneliese L. Sitarz
Dr. Robert J. Slater*
Dr. Cassandra Lynn Smith
Dr. Frank R. Smith
Dr. James P. Smith
Dr. Elizabeth M. Smithwick
Dr. Rosemary Soave
Dr. Richard L. Soffer
Dr. John A. Sogn
Dr. Leon Sokoloff*
Dr. Samuel Solomon
Dr. Martin Sonenberg
Dr. Thomas S. Soper
Dr. Hamilton Southworth*
Dr. Paul W. Spear*
Dr. A. Spector
Dr. Frank C. Spencer
Dr. D.B. Sprinson*
Dr. P.R. Srinivasan
Dr. Lisa F. Staiano-Coico
Dr. E. Richard Stanley
Dr. Neal H. Steigbigel
Dr. Charles R. Steinman
Dr. Ralph M. Steinman
Dr. Kurt H. Stenzel
Dr. Marvin Stern*
Dr. Stephen Sternberg
Dr. Irmin Sternlieb
Dr. John M. Stewart
Dr. Peter E. Stokes
Dr. William T. Stubenbord
Dr. Horace W. Stunkard*
Dr. Barnet M. Sultzer
Dr. Martin I. Surks
Dr. Roy C. Swingle
Dr. Marguerite P. Sykes*
Dr. John Taggart*

*Life Member.

DR. IGOR TAMM*
DR. J.R. TATA
DR. HARRY TAUBE
DR. SHELDON B. TAUBMAN
DR. HOWARD C. TAYLOR, JR.*
DR. ALVIN TEIRSTEIN
DR. HENRY M. THOMAS III
DR. DAVID D. THOMPSON
DR. G. JEANETTE THORBECKE
DR. NEILS A. THORN*
DR. DAVID A. TICE
DR. WILLIAM A. TRIBEL*
DR. WALTER TROLL
DR. JIR SHIONG TSAI
DR. GERARD M. TURINO
DR. KOJI UCHIZONO
DR. HIROSHI UENO
DR. JONATHAN W. UHR
DR. JOHN E. ULTMANN*
DR. FRED T. VALENTINE
DR. IVO VAN DE RIJN
DR. WILLIAM VAN DER KLOOT
DR. PARKER VANAMEE*
DR. ELLIOT S. VESELL
DR. MOGENS VOLKERT*
DR. SALMONE G. WAELSCH
DR. LILA A. WALLIS*
DR. GEORGE E. WANTZ*
DR. JONATHAN R. WARNER
DR. PAUL M. WASSARMAN
DR. LOUIS WASSERMAN*
DR. SAMUEL WAXMAN
DR. RENE WEGRIA*

DR. RICHARD WEIL III
DR. HAREL WEINSTEIN
DR. I.B. WEINSTEIN
DR. GERSON J. WEISS
DR. HARVEY J. WEISS
DR. HERBERT WEISSBACH
DR. GERALD WEISSMANN
DR. DANIEL WELLNER
DR. JOSEPH P. WHALEN
DR. ABRAHAM G. WHITE*
DR. M. HENRY WILLIAMS
DR. JOHN E. WILSON*
DR. MYRON WINICK
DR. JONATHAN B. WITTENBERG
DR. MURRAY WITTNER
DR. DAVID J. WOLF
DR. STEWART C. WOLF, JR.*
DR. JOEL A. WOLK
DR. HENRY N. WOOD*
DR. WALTER WOSILAIT*
DR. MELVIN D. YAHR
DR. MARTIN YARMUSH
DR. SEIICHI YASUMURA
DR. PETER I. YI
DR. BRUCE YOUNG
DR. MICHAEL W. YOUNG
DR. FULI YU
DR. RALPH ZALUSKY
DR. VRATISLAV ZBUZEK
DR. NORTON D. ZINDER
DR. JOSEPH ZUBIN*
DR. MARJORIE B. ZUCKER*
DR. DOROTHEA ZUCKER-FRANKLIN

*Life Member.